THE FRAMED WORLD

New Directions in Tourism Analysis

Series Editor: Dimitri Ioannides, Missouri State University, USA

Although tourism is becoming increasingly popular as both a taught subject and an area for empirical investigation, the theoretical underpinnings of many approaches have tended to be eclectic and somewhat underdeveloped. However, recent developments indicate that the field of tourism studies is beginning to develop in a more theoretically informed manner, but this has not yet been matched by current publications.

The aim of this series is to fill this gap with high quality monographs or edited collections that seek to develop tourism analysis at both theoretical and substantive levels using approaches which are broadly derived from allied social science disciplines such as Sociology, Social Anthropology, Human and Social Geography, and Cultural Studies. As tourism studies covers a wide range of activities and sub fields, certain areas such as Hospitality Management and Business, which are already well provided for, would be excluded. The series will therefore fill a gap in the current overall pattern of publication.

Suggested themes to be covered by the series, either singly or in combination, include—consumption; cultural change; development; gender; globalisation; political economy; social theory; sustainability.

The Framed World
Tourism, Tourists and Photography

Edited by
MIKE ROBINSON AND DAVID PICARD
Leeds Metropolitan University, UK

ASHGATE

Published by
Ashgate Publishing Limited
Wey Court East
Union Road
Farnham
Surrey, GU9 7PT
England

Ashgate Publishing Company
Suite 420
101 Cherry Street
Burlington
VT 05401-4405
USA

www.ashgate.com

British Library Cataloguing in Publication Data
The framed world : tourism, tourists and photography. --
   (New directions in tourism analysis)
   1. Travel photography--Philosophy. 2. Travel photography--
   Psychological aspects.
   I. Series II. Robinson, Mike, 1960- III. Picard, David.
   778.9'991001-dc22

Library of Congress Cataloging-in-Publication Data
The framed world : tourism, tourists, and photography / [edited] by Mike Robinson and David Picard.
      p. cm. -- (New directions in tourism analysis)
   Includes index.
   ISBN 978-0-7546-7368-2
   1. Tourism. 2. Travel photography. 3. Tourists--Psychology. I. Robinson, Mike. II. Picard, David.

   G155.A1F73 2009
   306.4'819--dc22

                                                                                2009006766

ISBN 978 0 7546 7368 2

Mixed Sources
Product group from well-managed
forests and other controlled sources
www.fsc.org Cert no. SA-COC-1565
© 1996 Forest Stewardship Council

Printed and bound in Great Britain by
MPG Books Ltd, Bodmin, Cornwall.

Contents

List of Figures

List of Contributors

Patricia C. Albers, Department of American Indian Studies, University of Minnesota, Minneapolis, United States of America.

Jenny Blain, Division of Applied Social Sciences, Sheffield Hallam University, Sheffield, United Kingdom.

Elisabeth Brandin, School of Human Sciences, University of Kalmar, Kalmar, Sweden.

Brian Cohen, Independent scholar, Pittsburgh, Pennsylvania, United States of America.

Teresa E.P. Delfín, Center for Latin American Studies, Stanford University, United States of America.

Stan Frankland, Department of Social Anthropology, University of St Andrews, St Andrews, United Kingdom.

Janet Hoskins, Department of Anthropology, University of Southern California, Los Angeles, United States of America.

Joyce Hsiu-yen Yeh, Department of Indigenous Cultures, College of Indigenous Studies, National Dong Hwa University, Hualien, Taiwan.

Vassiliki Lalioti, Department of Music Studies, National and Kapodistrian University of Athen, Ilissia, Greece.

Marie-Françoise Lanfant, URESTI, Centre National de Recherche Scientifique (CNRS), Paris, France.

Andy Letcher, Department of Theology and Religious Studies, Oxford Brookes University, Oxford, United Kingdom.

Ilyssa Manspeizer, Independent scholar, Pittsburgh, Pennsylvania, United States of America.

Matthew J. Martinez, Departments of American Studies and American Indian Studies, University of Minnesota, Minneapolis, United States of America.

David Picard, Centre for Tourism and Cultural Change, Leeds Metropolitan University, United Kingdom.

Celmara Pocock, Queensland Museum, South Brisbane, Australia.

Mike Robinson, Centre for Tourism and Cultural Change, Leeds Metropolitan University, United Kingdom.

Rebekah Sobel, University of Maryland, College Park, United States of America.

Robert J. Wallis, Richmond University, London, United Kingdom.

Elvi Whittaker, Department of Anthropology, University of British Columbia, Vancouver, Canada.

Acknowledgements

This book was conceived as a collection of perspectives on the relationships between photography and tourism. We sought to bring together contributors from different disciplinary perspectives and traditions who were intrigued by the central role that photographs and photography play in the lives of tourists and the phenomenon of tourism.

In the course of bringing these perspectives together some of our contributors have moved, married, changed jobs and generally grown older. As editors we have perhaps grown much older than everyone else! Being philosophical about it, we'd like to think that the book has matured over the time taken to bring it publication. We would like to extend a warm thank you to all of our contributors for their patience and good humour in putting this volume together. We believe that all has been worth the wait despite our grey hair.

Special thanks go to our colleagues Philip Long, Simone Abram, Alison Phipps for their advice and assistance along the way, to Konstantinos Arvanitis, and to the various reviewers who have overseen the project and have offered their helpful and constructive advice.

Thanks also to Ashgate Publishing, for their professionalism, patience and dedication.

Mike Robinson
David Picard
Centre for Tourism and Cultural Change,
Leeds, 2009

Chapter 1

Moments, Magic and Memories: Photographing Tourists, Tourist Photographs and Making Worlds

Mike Robinson and David Picard

Introduction

To be a tourist, it would seem, involves taking photographs. Whilst photography is clearly not the exclusive preserve of tourists it nonetheless is one of the markers of *being* a tourist (Markwell 1997), is intimately linked to the *doing* and *performing* of tourism (Bærenholdt et al. 2004) and, is indeed a constituent element of the tourism industry as it works and plays with the signs and images of visual culture that is everyday life (Mirzoeff 1998) in order to project parts of the world onto other parts of the world. Susan Sontag, writing over thirty years ago, highlights the apparently normative and accepted relationship between tourists and photography and comments that; 'it seems positively unnatural to travel for pleasure without taking a camera along' (1977, p. 9). More precisely, the obvious presence of a camera about the person has tended to delineate someone as a tourist, or rather as someone who is equipped as a tourist; primed to capture images of a different place (Edensor 1998). As John Hutnyk (1996, p. 145) notes: 'Holiday photography is the record which shows, no matter how rushed the visit, that what was seen was what was there'. This notion of recording encompasses a variety of practices and rituals in itself, but it extends beyond ideas of collection and record and into the realms of self-making, authentication and socialization processes which are bound up with the embodied doing of tourism (Crouch 2000, 2002; Crouch and Lübbren 2003). The holiday photograph evidences not just the tourist site but the tourist him/herself, in a form of expressive self-creation (Crang 1999).

Stereotypes of American and Japanese tourists conspicuously bearing the latest photo-technology around their necks are still prevalent within popular depictions of 'holiday-makers' and despite any emergent reflexivities amongst tourists, and for toured communities, the carrying of a camera still signifies an 'outsider'. Orvar Lofgren (1999, p. 82) comments: 'The fact that a dangling camera has become the sign of the vulgar tourist poses a problem for those who feel a need to distance themselves: should they carry a camera at all?' In line with contemporary digital technologies and the advent of photo-mobile phones, cameras are now widely carried in the course of daily life; we are almost perpetually primed to click.

However, it is within vacation spaces and times when cameras are still at their most visible and indispensable. The tourist *becomes* a photographer in an instant. Craft, training and technological prowess of photographing are quickly learnt, or indeed bypassed, as the sense of the moment takes over. Sometimes spontaneous, sometimes staged, the being in a different environment and 'on holiday', not only presents opportunities for taking large numbers of photographs, but also for the full range of rituals which accompany tourist photography before, during and after the trip.

Underlying the momentary and apparently frivolous act of taking photographs while on holiday are a host of searching questions which link to the ways in which tourists experience and negotiate the world and the ways in which the world regards tourism. Why *do* tourists take photographs? Why do tourists frame certain things, while other things remain outside of the frame? *How* do tourists photograph the Eiffel Tower or Native American Chiefs and the vast variety of peoples, places and events which they encounter? What does it *mean* when tourists 'shoot', 'capture' or 'take' photographs? How does tourism photography mediate or change the realm of the photographed? Who controls tourist photography and who 'owns' the photographic image? Do tourist photographs need captions or 'voice-overs'? What happens to tourist photographs once tourists have returned home? What do tourist photographs tell us about the relation between the photographer and the photographed? This list of questions is far from being exhaustive and spills over into wider and fundamental issues relating to visual culture and its dominance within the modern world; issues such as those relating to power, authority, aesthetics, play and individual/collective identity. The tourism industry (comprising tour operators, airlines, accommodation providers and destination marketing authorities etc.) has long been versed in presenting the world and its peoples through photographs, and in doing so it implicates itself in the construction of influential narratives which reach beyond the sphere of tourism and into generic narratives of representation. Understanding the relationship between tourism, tourists and photography is a way of approaching some of the most vivacious theoretical, epistemological and ethical debates taking place within the field of tourism studies over the past thirty years.

Bringing the World Home

Travel and exploration as the great projects of the Renaissance relied upon ways of recording the world. Discovery, conquest and encounter were not only based upon the physical acts of travel, battle and trade but significantly upon the representations and circulations of these acts to the vast majority of populations who essentially remained 'at home'. Here was the educational function of travel as mediated through the texts and images produced 'in situ' and brought back for conspicuous display. While journal narratives and diaries were important, it was the visual recordings of destinations and peoples which provided the greatest impact of the various forays into an expanding world, albeit amongst a rather narrowly defined social elite. Watercolour and pencil sketches from both professional and amateur artists, captured an emerging world; a myriad of sites and sights, some of which had been described first, second or third hand, and others which were truly 'new' and also those which were wholly imagined. Paintings in the vein of the classical and sublime aesthetic and the emergent picturesque, of the ruins of Athens and Rome, or of landscapes and natural features, were displayed in the galleries and libraries of leading houses, integrated to their gardens, and were reproduced as woodcuts, engravings and lithographs in books. Though the landscapes, objects and events recorded through the grand tours of Europe and explorations beyond, fashioned, and in some cases, accentuated tastes, this was more than a subjective artistic endeavour; it constituted attempts to 'capture', order and curate the world in the very process of it being discovered (Boorstin 1985) and introduced the earliest semblances of a visual culture as something not only participated in and understood by individuals, but as something circulated and shared through society.

Judith Adler in her consideration of the 'Origins of Sightseeing' (1998) has marked out the historical context for photography as a convergence between post-Renaissance science and travel, and the shift to objectification and the recording of 'things'. In the ferment of secular travel and the discovery of 'new' topographies from the sixteenth century onwards, the privileging of sight and of seeing was a way in which travellers could, in Adler's terms, 'grasp this vast new world of things without being overwhelmed by it' (p. 19). Despite the scientific reductionism inherent in the visual capture of the world, the mediation of this early gaze was via the paintbrush and pencil and the objects created were essentially works of art and thus still laden with the ambiguities of the imagination. They also required an educated 'eye' to appreciate them, something easily at hand since the audience for such representations of places near, and increasingly far, remained erudite social elite. The world was travelled, recorded and appreciated by a relatively small social grouping.

The introduction of photography in the late 1830s demonstrated several continuities but some important dis-junctures with the pictorial representations of an expanding world. The scientific ethos of objectifying the world was accentuated in that the interpretive veil of artistic representation was removed to produce novel realities. Though photographs were subject to manipulation, and a certain aesthetic continuity with paintings and prints in terms of style and framing, to all intents and purposes they allowed their audiences a far more direct way of seeing the world. In the early years of photography the audience for photographs remained as relatively small and select as it was for the paintings and sketchbooks of the upper classes. Significantly however, throughout the nineteenth century photography generated substantive new audiences amongst the growing middle classes. The fascination with the science of photography as a way of reducing and ordering the world was quickly overtaken by the practicalities of technology which was much more concerned with, and allowed, the rapid diffusion of images of the world.

Photographic exhibitions which thrived during the nineteenth century were seldom viewed as any sort of artistic endeavour but rather as visual encyclopaedias which reflected the technological prowess behind transport developments and, through the frequent portrayal of places, also reflected colonial dominance.

It was the development of a new visual economy which saw the photographers selling negatives to libraries and commercial printers and publishers, which in turn distributed to an audience eager to consume the records of travel to distant lands. Photography by travellers allowed a relatively small number of individuals and groups to appropriate and moderate ideas of the diversity and difference in the world and essentially control an economy of images. But the boundaries of this economy were not easily fixed nor controlled. The visual is too appealing to be restricted. Photographs *of* the world spilled *into* the world through various forms of reproduction. As Peter Osbourne (2000) explains it:

> The immediate application of photography to the depiction of travel is explained by the fact that it was, on the one hand, a crystallization of three hundred years of culture and science preoccupied with space and mobility and, on the other, the expressions of its own time – the epoch of capitalist globalisation, the construction of a new middle-class identity and the dramatic speeding-up of transportation and communication. Photography was a representational tool refined in the service of these processes. (p. 9)

Developments in photography since its discovery in the late 1830s have closely tracked developments in transport technologies and generally reflect processes of massification and ever-increasing circulations of knowledge about the world. Of course in the initial stages of development, photography as a practice was extremely limited in a very real sense as the equipment needed was not only expensive it was extremely cumbersome (Gautrand 1998). But for those travellers and explorers who did endure the trials of carrying around vast quantities of photographic equipment the rewards were great. The convergence of commercialism and photography was a far swifter, deliberate and more obvious process than that between commercialism and art. When Francis Frith, for instance, travelled and photographed the Middle East in the 1850s he was already well aware of the growing number of tourists to the iconic centres of Egypt, Palestine and Syria (Wilson 1985). Though Frith never engaged again in foreign exploration, the returns from his travels were significant and came in the form of exhibitions, lecture tours and magazine articles. The history of nineteenth century travel photography is essentially grouped around a number of technologically minded artists and commercially minded technophiles such as Frith, each playing a part in the structuring of what was to become the tourism industry. Photography created an essential shop window for the world and became an essential pillar of modernity, not only in terms of its underlying practical technological advances, but also as the means of documenting discovery and on-going social and cultural changes.

The marriage between photography and travel was marked not only by the relative perfection in which a subject or landscape could be captured, but also by the notion of immediacy; the sheer speed of the process of transformation from reality to its 'accurate' representation, generally mirroring a 'speeding up' of social life and the closing down of distances. It is immediacy which more than anything has defined the evolution of photography, what Walter Benjamin (2006) referred to as the 'Here and Now', as if 'reality has, so to speak, seared the subject'. (p. 243)

Photography as Popular Culture and Instant Power?

Throughout the nineteenth and into the twentieth century, photography largely remained an activity of a relatively small number of producers and publishers. Prohibitive costs, bulky equipment and the dominance of a relatively small number of experienced photographers effectively controlled the circulation of images. Even tour operators such as Thomas Cook persisted in employing hand drawn illustrations in advertisements, posters and brochures well into the 1950s. The tourist of the early twentieth century would ostensibly rely upon the purchase of a photograph/postcard from an agent or distributor of a professional photographer. In destinations such as Egypt, Turkey and Greece tourists would be directed via guides and guidebooks to established photographers noted for their expertise, the quality of their studios and the extent of their views. Notably, these were generally Europeans who had travelled and settled in such countries as part of an active commercialization of 'the Orient' through the sale and distribution of photographs (Ryan 1997; Micklewright 2003). As Katrina Thomas (1975) notes:

> In Cairo, the Germans seem to have dominated the photographic scene from the first. In Baedeker's Egypt, Handbook for Travellers published in 1885, Schoefft, Stromeyer and Heymann are listed as the chief sellers of photographs, though there is also 'Sebah of Constantinople' and Laroche and Co. Schoefft is recommended for 'a good background for groups; also a fine collection of groups of natives and a few desert scenes, some of which are very striking'. (p. 27)

But while generally cheap and widely available, the photographs sold to tourists clearly reflected the world as a series of views, framed by the professional (often without leaving the studio) and his (very rarely 'hers'!), artistic sensibilities and technological prowess. In the context of the, then developed world, and despite the rapidly expanding population of tourists drawn from the middle classes, taking one's own photographs of a destination remained out of reach financially and in terms of technical and artistic skill, to a point where the development of a modern tourism as a social practice out-paced the technological developments of photography.

This was a situation which was not to last long and the *buying* of photographs gave way to a *taking* of photographs particularly as George Eastman, the founder of the American Company of Kodak, offered the 'Kodak Brownie' at a retail price of one dollar in 1900. This 'point and shoot' camera was a development of an earlier camera designed by Eastman and offered for sale in 1888 at the price of twenty dollars, along with the slogan: 'you press the button we do the rest'. Despite having to return the camera to the factory to develop the prints in its first year of production some 13,000 were sold (I'Anson 2000). The Kodak Brownie when first offered in 1900, complete with a film roll for 15 cents, shipped over 150,000 (see <www.kodak.com>). It is beyond the scope of this chapter to detail the history and development of the camera (see for instance: Willsberger 1977; Wade 1979; West 2000) suffice to say that in line with the language used to advertise cameras since the 'Brownie', all the individual had to do was to 'click' or 'snap'. This transfer of control and power from a relatively small group of 'experts' to an increasingly mobile mass was arguably the greatest single event in the shaping of tourist identity. While developments in 'public' transport technologies have clearly enabled people to travel, and 'destinations' to be designed, the camera as a *personal* object enhanced (in theory at least), not only the individualization of 'seeing the world', and situating ourselves in images of it (Berger 1972), but also allowed us to participate in the very construction of these images.

Pierre Bourdieu in his treatment of photography as a 'middle-brow art' (1990) and how it sits in terms of class structures and socio-cultural practices, recognized photography as a mass social practice rather than as something to be defined by any set of intrinsic qualities. Indeed, Bourdieu locates photography as a family matter and thus the occasions and spaces of family get-togethers (which of course includes the holiday) become the settings for 'playing' with the camera. The evocation of play is important here as it celebrates amateurism and the instantaneous. The holiday snap, featuring family or friends, communicates a very real sense of playfulness with people pulling faces, standing in ridiculous poses and consciously celebrating the disjuncture with normal work related behaviours.

The developments in technologies right up to the present with digital cameras and mobile phones as instruments of hyper-instantaneousness are refinements of a process of empowerment through which tourists are playfully able to create their own narratives of being elsewhere. In a historical sense, the ubiquity of camera ownership amongst tourists, and indeed amongst a majority of the population, could be said to demonstrate a transfer of 'power' from the preserve of relatively few early photographers. All tourists now possess the power to capture the world themselves. John Urry (1990, p. 138) comments that: 'To have visual knowledge of an object is in part to have power, even if only momentarily, over it'. Urry cites the example of the way in which photography has 'tamed' so-called 'exotic cultures'. But while this 'knowledge/power' relationship which Urry identifies as being a feature of tourist photography could be ascribed *post hoc* as being instrumental in, for instance, perpetuating colonialist relations, it would not seem to be consciously exercised by the tourists themselves. Unlike the corporate appropriation of imagery which does possess real power to represent/mis-represent, the pressing of a button, and the demonstrating of a kind of technological prowess by the holidaymaker would seem to be directed not at any outer exertion of power over the object, or 'other', but rather at the self. The millions of holiday snaps of peoples and places, while demonstrating the power of instantaneous capture are usually destined for dusty cupboards rather than any influential global circulation (arguably, the 'posting' of holiday photographs on the world wide web and as part of an increasing number of travel 'blogs' does technically allow a form of global circulation, but, the extent to which this demonstrates any influence on behaviour is questionable). Moreover, future interrogation of photographs by friends and family is frequently centred just as much upon the stories of the photographer as it is on the objects photographed (Holland 1991; Hirsch 1997, 1999).

Photography, through mass production and, critically, mass circulation has been central in accelerating the construction of 'place myths' (Shields, 1991) or 'imaginative geographies' of the world (Gregory 1995; Schwartz 1996; Schwartz and Ryan 2003; Gregory 2003); geographies that reflected both national and cultural identity (see for instance: Jager 2003; Snow 2008). Many of these inscriptions of place have changed little since they were captured in the nineteenth century. However, once photography shifted into the realms of the masses, and particularly once the masses *became* tourists, imaginative geographies became domesticated (Chambers 2003) and images of the world were imbued with private, family meanings and personal narratives. This has entailed the creation of a parallel, yet joined system of representation. On the one hand, widely circulated 'public' (and largely professionally taken) photographs fuel the construction of the world as a 'tourist world' of brochure and guidebook destinations, working closely as they do with universal narratives of exoticism, versions of paradise and the spectacular (Goss 1993; Cohen 1995). On the other hand, images of the world are also held in myriad private collections as private (and largely amateur) photographs, sporadically annotated and each reflecting very different and intimate experiences, imaginings and meanings of the world. Each set of visual representations exist, and are connected; the images of a Caribbean holiday beach strewn massively across an advertising hoarding and demonstrative of the super-structures of the international tourism industry, are reflected in, and animated by, the photographs of the family on holiday on a similar beach (Albers and James 1988; Urry 1990), which display not only the shadows of the present world but which also reproduce accumulated histories and epistemologies. As Bourdieu (1990, p. 75) notes: 'The ordinary photographer takes the world as he or she sees it, i.e. according to the logic of a vision of the world which borrows its categories and its canons from the arts of the past'. The photographs of tourists, Olivia Jenkins (2003, p. 308) points out, are part of the hermeneutic 'circle of representation'. In this sense, the 'holiday snap' is a moment of revelation on the global power of the visual and is implicated, often unknowingly, in a connecting to, and the creation of the 'outside' world and the 'other'.

The Spaces and Magic of 'Capture'

With the advent of the widespread social ownership of cameras also came the end of a dependency upon the 'professional eye' and an *apparent* liberation from an artistic, and acutely romantic, 'expert' framing of the world. Camera owners were/are able to 'capture' the world, or rather moments of being in the world, notably whilst on vacation. The context of the vacation as an intersection of space and time for the masses rather than an elite few, is critically important and yet rather peculiarly underplayed in much of the literature relating to photography as a social practice. Examination of photographs as historical record, as conveyors of messages and as artwork, together with various readings of psychoanalytic and cultural signs contained within the images, have drawn heavily on 'professionally' produced photographs, many from the first century of photography, and on public photographs which have been widely reproduced in books or as postcards. There has been a general lack of analysis of holiday photographs despite the fact, as Patricia Holland (2000, p. 148) notes, that 'most collections of personal pictures are, in fact, dominated by time spent *away* from home'. Travel photography, essentially that which was produced by the pioneers of both travel and photography, or that which makes the pages of today's *National Geographic Magazine*, continues to be a strong source of material for interrogation and mirrors the persistent, if specious distinction between tourist and traveller (Strain 2003).

Professional travel photography constitutes its own, if very broad, genre which borrows strongly from the techniques of the artist. Perspective, light, juxtaposition between foreground and background, together with various textures produced in the chemical development/production of the photograph reflects a technical process of order and composition. In the context of travel photography, destinations, and their inhabitants, are invariably artistically framed. The expert intervention of the photographer composes and creates. The photogenic is that which invites artistic composition by virtue of its fit with preconceived notions of artistry to the point of the stereotypical (Gombrich 1960), with 'views' being *constructed* and replicated, based upon accumulated knowledges and culturally located ideas of what is attractive. MacCannell (1976) refers to this as 'sight sacrilization'; the process of inscribing sights through photography and the further circulation of these images to the point where they become almost the only markers for the sight. Deviations from normative constructions of what is usually considered to be attractive for a photograph are themselves frequently aestheticized and imbued with an artistic quality. Photographs which, for instance, capture industrial decay, poverty and sites/sights of trauma, can slip into the fuzzy realms of 'art', making them at once almost desirable places to visit (Edensor 2005).

The professional travel photograph is woven closely into the structures and presence of international tourism, and in the consciousness of international tourists. These are the photographs which appear on postcards, in magazines, guide books and brochures. The professionalism and the technique behind such photographs marks them out as carrying both an economic as well as a symbolic value, and characteristically they have their own market and are bought, sold and exhibited as highly public commodities. The iconic photograph which portrays a place and/or its people, as the work of a professional photographer, technician or indeed, an ethnographer, usually has a long lifespan and is able to circulate and permeate collective consciousness. These are the photographs that are reproduced, copied, mimicked, and parodied and consequently have tended to attract the attention of researchers as the objects of study.

In contrast to the canon of travel photographs, vernacular tourist photographs and holiday 'snaps' have attracted limited attention from the research community. Despite the incalculable number of such photographs in existence they are, by definition, located largely in the private rather than public sphere and consequently removed from scholarly consideration. Arguably, and linked to their public absence, their social impact is somewhat minimal; visibility being all. Regardless of various degrees of competency and artistic flair with which the photographer may capture the occasion of travel, and the holiday experience as a series of frames, the process of photography is divested of technical reference points and the holiday photograph is almost entirely an amateur object. The tourist photograph makes no claims towards art and is largely removed from the worries around 'art as photography' as expressed by Walter Benjamin (1999). The artistic emptiness (or at least a shallow mimesis of the artistic), amateurism, and the apparent 'non-serious' nature of contemporary tourist photographs, along with their public absence, together with a faint aroma of snobbery on the part of technicians, artists and some academics, has resulted in their omission from sustained analysis. At one level this is understandable and it would make sense to focus attention on photography which has widespread influence upon the communicative networks within which we function. Even Roland Barthes, who in *Camera Lucida,* recognised the problems of classifying/reducing photography and the need to adopt a more primordial approach to effects of all photography on the spectator/reader, chooses to reflect upon as his examples, decidedly public photographs; usually by known and named photographers. These are the photographs which reach us, not only emotionally, but physically. These are the photographs we are confronted with daily; part of the world's 'generalized image-repertoire'. (Barthes 1993, p. 118)

On another level, in privileging the widely seen images of the few, we neglect (and in some ways resist), the widely experienced world captured by the many. The epistemological problems relating to the private photographs of others, and to the photographing by the masses of tourists, are not in themselves enough to jettison either their intimate meanings or their wider social importance. The albums of family holiday photographs are testimony to the transfer of the 'magic' of capture (Picard, forthcoming); a sort of 'technology of enchantment' (Gell 2005), which has passed from the expert to the amateur, in parallel to the dissolution of categories of traveller and tourist.

The term 'capture' is telling. Not only does it invoke notions of ownership, it also carries meanings of order and structure. But what precisely is captured in tourist photography? Before answering this question, it is necessary to analyse the cultural and ontological premises underlying photography as part of tourist practice. A defining element of tourism lies in the distance between holiday destinations and the everyday life spaces of tourists (MacCannell 1976). Tourism mobilizes people to travel to, and through, places that are geographically separated from their homes. Distance and geographical separation are understood here as socially constructed notions that emerge from specific historic contexts framing, and consequently structuring the subjective perception of people's 'being in the world'. They can be understood as representational systems embedding the uncertainty of social, spatial and temporal distance, extremes of infinitude and absence, the chaos of quotidian events and happenings, the inner turmoil of desires, emotions and dreams within forms of aesthetic and moral order.

In the context of tourism and the wider realms of modernity from which it emanates, various media, including tourism brochures, guide books, travel forums together with future tourists in preparation of their trips themselves, appear to model their ideas about 'destinations' according to a largely collective imagining (Crouch, Jackson and Thompson 2005). Destinations are constituted as, more or less, closed systems; as intrinsically structured worlds built upon sets of aesthetically significant sights. The geographical separation and distance with these 'worlds' mark a more important ontological separation and distance where destinations are enchanted and 'othered'. The touristic journey to, and through, these 'other' lands and back home, can be seen both as an existentialist reflection upon life and, as a reproduction of power structures relating to the very perception of distance at a wider sociological and political level.

While on holiday, the distance between the spaces of everyday life and the quasi-mythical holiday destination decreases and ultimately disappears. The places visited and the peoples that inhabit them become very real. The ontological separation defining tourists' home spaces and holiday destinations consequently appears reduced, indeed, nullified, during what Stephanie Hom-Cary (2004) terms 'tourist moments'. These are the brief instances during which tourists connect to the 'other'; moments of social intimacy during which feelings of difference are inverted into feelings of sameness, of being part of the same cosmic condition. The tourist's confrontation with people, spaces or stories is not only a confrontation with 'others', but maybe more importantly, a confrontation to possible selves.

In encountering the 'other', tourists are provided with opportunities to recognise and confront the persons that they are themselves, which they were before, or will be in the future, or have never dared to be. The coming together of peoples and place, of projected and imagined selves within different landscapes and languages, in moments of pleasure and angst, are all dimensions of the touristic experience. The physical place in which tourism unfolds – the 'destination'– seems to work simultaneously as a giant theatre, a giant theme park, and a giant Freudian couch, confronting tourists to various personas they think that they are, they desire or imagine to be or, wish to become. In this sense, the collective frame of tourism practice and of its related production of sites and experience clusters in destinations allows tourists to connect on a very personal level to multiple facets of their identity; to reflect upon and ultimately recreate their being in the world.

The magic of the camera, the enchantment of photographic technology (Gell 2005) may precisely lie in its believed capacity to capture the spirit of a place, to create copies imbued with the power of the original place visited. In this sense, the camera is seen as a technology able to transport some essence of a place into another without losing its material authenticity. The technology of the camera is based on the framing and breaking of light through an optical system and the projection of this processed light onto the surface of film or, more recently, an electronic chip. In the traditional technology the more or less intensive light induces a chemical process which transforms the surface of the roll of film and records differences in light intensity. With digital technology, this recording is made via a large number of evenly distributed light sensitive sensors which transform the optical image into digital information. At a later stage, photographic prints are developed through photo-chemical processes or a high definition printer or electronic screen. In both analogous and digital photography, the physical contact between the recording device (the film or chip) and the photographed realm is established through light.

One could argue that, in analogy with Renaissance techniques of capturing the power of relics by using a mirror, light continues to be conceived as a possible medium to transport specific qualities attributed to specific realms; as a cause for contagion bringing separated worlds in physical contact. In this sense, the power of photography to convincingly record and mediate the 'essence' and 'inherent truth' of the photographed (Sontag 1977, Taussig 1987) may lie less in its technology of mimesis – the technology of copying – but in its technology of trans-materialization. It may lie in what Susan Steward (1993) called the 'authentic link' established between the photographed object and the final print; a link that convincingly authenticates the truthfulness of the photography. Tourists usually prefer taking photos with their own cameras, rather than buying postcards which are normally reserved for sending to others. They seem to need to establish the technological act of photographing themselves. The initial worries expressed by many amateur photographers that the recently popularized digital photography is somehow not 'real' photography further affirms this idea as if real' photography needs 'physical' contact to generate a genuine copy.

At the same time, this argument is today largely being challenged by the very success of digital technology in the photographic mass market. Digital cameras are based on the translation of differences in light intensity into a digital electronic signal. In analogy to the shift in sound technology or, much older, in pharmacology (copying the effect-relevant surface structure of molecules), the 'physical' chain between the original and the 'legitimate' copy hence seems broken. Yet the popular use and the apparently continuous meaningfulness of copies obtained through digital technology, as witnessed by the growth of digital tourism photography during the first years of the twenty first century, seems to indicate an ontological or, at least, a semantic change. Digital signals transmitted through an electric current seem to have been legitimized as 'fluids' (Urry 2003), that can transmit authentic

Figure 1.1   Instant capture of the screen

Note: In St Stephen's Cathedral, Vienna, a tourist takes photographs of computer images of those parts of the Cathedral not open to the public on that day.

'magic power' in the same way as other forms of imagined or real physical contact in other times. As part of a call for re-thinking our conceptual and methodological frameworks for examining the photography of the digital camera and its related modes of storage, communication and the new social networks it is producing, Jonas Larsen (2008) makes the observation that while analogue photography was focussed on future audiences, the photography of digital cameras and mobile phones is directed towards an immediate audience. While this is only partly the case, as prints are still made and shared and even computer stored images are recalled for the future, it nevertheless points to a need for re-assessing the context of taking photographs. John Tomlinson (2007) observes that digital photography is only one part of a plethora of social practices of modernity which are symptomatic of what he refers to as the 'culture of instantaneity'; increasingly a normative state which is now familiar with, and expectant of, 'rapid delivery, ubiquitous availability and the instant gratification of desires' (p. 74), the dominance of the digital camera would certainly seem to have increased the technological capacity of tourists to capture the world (Figure 1.1). However, the practices and the meanings of 'doing' photography while on holiday display a comforting continuity.

Figure 1.2   A typical tourist photograph of the Treasury of Petra, Jordan, 'naturally' framed by the pathway through the landscape

Framing and Performance

Tourist photography is littered with performances involving those taking the photographs, those 'in' the photographs, those outside of the frame and, all those who will later work with the photograph itself as a catalyst for storytelling.

The taking of photographs involves a framing of the world; a procedure of focus, both literally and metaphorically (Figure 1.2).

The subject, be it a person or a landscape, is selected and necessarily, other people and parts of the landscape are excluded. Framing is an ambiguous practice. On the one hand, the act of framing removes a sense of context (Duro 1996) and the resultant image is disconnected from the wider landscape or chronology of life events. In taking a photograph of a monument or a festival, for instance, the tourist can be seen to be 'liberating' the moment and essentially making such moments somehow 'objective' (Terdiman 1993). This objectifying

Figure 1.3   Tourists queue and cram into a small space to photograph the mosaics in the Hagia Sophia Mosque, Istanbul

can be seen as a way of privileging or reifying what is shown. On the other hand however, the photograph, despite its clinical cutting out is seldom left to speak for itself, or indeed, speak at all. In counter-point to professional and 'official' images which circulate on bill boards and in magazines, the person taking the photograph, maybe combined with others behind the camera, or indeed those 'in shot', normally provides a commentary for the image. This frequently involves a post-rationale for the photograph such as; 'in this one I was trying to show … ' Connections and contexts can be re-established as the photograph and its object is re-made through narrative and its placement in (at the very least, family) history (Berger 1980). Of course, such stories are bound up in our own subjectivities, but then again so is tourism. So whilst reification is part and parcel of framing the tourist photograph it is not necessarily condemned to be forgotten as Adorno and Horkheimer (1997) would have it.

The act of framing the holiday photograph is part of a performance which in itself has become part of the wider 'doing' of tourism. This is very much an amateur performance though it may well mimic the performance of the professional. However, as Vilem Flusser (2007, p. 58) observes:

Figure 1.4   Tourists gather around a bronze statue in a Budapest park to play and pose for photographs

> People taking snaps are distinguishable from photographers by the pleasure they take in the structural complexity of their plaything. Unlike photographers and chess-players they do not look for 'new moves', for information, for the improbable, but wish to make their functioning simpler and simpler by means of more and more perfect automation. Though impenetrable to them, the automaticity of the camera intoxicates them.

In the practice of 'framing' their photographs, tourists enact certain aesthetic preferences and draw upon wider representations, knowledges and histories of a location. Tourists jostle amongst themselves for positions which will allow them to capture the 'best' view or clearest perspective (Figure 1.3).

Julia Harrison (2003, p. 100) describes how tourists have often elaborate strategies for achieving a 'good' photograph. Writing of one tourist she records:

> Michael often emphasised the formal compositional dimensions of scenes that he wanted to photograph during his travels. He showed me several photographs of doorways with people standing in them, a framing device that helped him 'get a better picture', bring together colour and composition. As he said, 'Another trick I like for a good picture is that I walk down the street and I say – "Oh, isn't

that a nice doorway. But, it would be more interesting if I had a person in it". So,
I wait till somebody walks into / or through the doorway'.

Queues of tourists can be identified waiting to take the 'classic' shot of a building
or landscape. Common conventions played out by tourists include lining up objects
in as 'near as perfect' symmetrical ways, or waiting until all other tourists are out
of view to give the impression of temporary sole ownership of a sight. Tourists
observe other tourists taking photographs and in a fit of mimetic contagion can
seek to do the same (Figure 1.4).

Some tourists follow the signs often provided which cue them to 'take
photograph here', or at least denote a 'scenic view point'. Other tourists, in acts
of resistance look for different angles. Such behaviour, parodied wonderfully by
the work of Martin Parr (1995), while mildly amusing in retrospect is common
practice by many and is indicative of the ways in which tourists are shaped by the
dominant aesthetic tropes which circulate through the world.

Though clearly no statistics are available, holiday photographs would appear to
be easily divided into those which *do not* feature friends and family and consist of
'neutral' views of landscapes and buildings, and those that *do* foreground friends
and family. It is the latter category which engenders various ritual performances
from both photographer and subjects. Groups are lined up and ordered by the
photographer who takes on a position of temporary authority; shortest in front,
tallest behind, girls one side, boys on the other, etc. Such momentary curations
are indicative of the seriousness behind a fleeting and seemingly frivolous act.
Getting everyone in the shot and not leaving anyone out of the frame entails
direction and responsibility that itself is subject to discussion, negotiation and, at
times mediation. Traditions of 'saying cheese' (or equivalents) as a way of getting
subjects to smile are played out across cultures (Kotchemidova 2005).

For subjects in the frame, various performances are played out, involving a
range of solo and group bodily poses and facial expressions. As Michael Haldrup
and Jonas Larsen (2006, p. 283) put it:

> Tourist photography is part of the 'theatre' that enables modern people to enact
> and produce their desired togetherness, wholeness and intimacy. When cameras
> appear, activities are put on hold, and in posing people present themselves as a
> desired future memory; they assume tender postures: holding hands, hugging,
> embracing and so on.

While family group photographs can be traditionally orchestrated to produce
a uniform smiling line of subjects, emphasis is frequently on play and 'out
of character' behaviour, often involving 'props' – physical objects that are
highlighted and used by the subjects to accentuate the situation (Latour 1991). The
dialectic produced between the materiality of location and the posing participants
of photographing not only feeds play but can contribute to the formulation of
emotional geographies (Hallman, Mary and Benbow 2007) anchored amongst

Figure 1.5   **One of many 'backdrops' at Disneyland, Paris complete with a mobile 'prop'**

family and friends, so that backdrops become tied to the emotions being explored and 'built' through photography.

In the earlier days of photography, the professional studio could usually provide a variety of props and backdrops as aids and substitutes for reality, allowing anyone to be captured as a tourist without having to travel. Stage sets in variety were also brought on streets in resorts and towns in order that tourists could be captured in full fantasy and pose against, or next to, the everyday (Lippard 1997) or as and an extension of reality (Sandle 2003). However, the landscape or streetscape provides the tourist with obvious background for any performance and can be manipulated by subjects for effect as if it were a prop. Holding up the leaning tower of Pisa, or showing the Eiffel Tower as emerging out of someone's head, have in themselves become ritual poses for the holiday snap and typify a playfulness and an amateurishness, and to some extent the spontaneity of the occasion. But while holiday location becomes an immediate and convenient canvas for the tourist, backdrops are seldom neutral or passive. Arjun Appadurai (1997) observes that they locate the photograph in wider public discourse, frequently the discourse of modernity reflected in the clothes and attachments of the tourist and often juxtaposed against the social subaltern. This discourse is accentuated in the context of the postcolonial and as Appadurai argues: 'photographic backdrops and props play an increasing role in the work of the imagination, in consumer-driven images of subjectivity and in socially mobile practices of self-representation and class-identification'.

What are essentially spontaneous embodied performances amongst tourists themselves (Schechner 1988; Turner 1990), both behind and in front of the camera, are provided by, and constructed from, the intersections with various backdrops (Figure 1.4).

This is, more or less, common across all photography, but in the context of tourism the backdrops and relations with them take on the added significance of ephemerality and, unlike that of one's own garden or living room, are largely outside of the control of the performers, so disruption may occur (locals, the subaltern, other tourists can invade the frame and distort the backdrop), or, the backdrop may not match up to that provided by the tourist brochure. Despite these features of tourist photography, commonalities and continuities with the everydayness of photographic practice and performance are clearly discernible. As Steve Garlick (2002, p. 5) puts it: 'We cannot separate "tourism" off from the broader social and cultural processes that determine its forms'. Key to the performances produced for, and to an extent, *by* the camera are the socialities it both displays and engenders. Such performances are underpinned by both extant and newly created group dynamics which revolve around family and intimate social relations (Larsen 2005). As Haldrup and Larsen (2003, p. 42) argue in one the few research articles which address the 'doing' of holiday photography (in their research focusing upon Danish tourists), the performances that are enacted demonstrate that 'tourist photography is closely tied up with staging social relations and transforming places into private theatres of blissful family life'. They continue to reflect that:

> ... photography and tourism are major social practices through which modern people produce storied biographies and memories that provide sense to their selves and their social relations. Tourist photography and family photography are wedded thoroughly into each other, which again reflects that in a modern world organized mobility, tourism is not an 'exotic island' but connected into 'ordinary' social life. (p. 42)

Such observations feed the thesis of tourism as a set of performances which blur with those of the everyday. The holiday in this sense provides yet another occasion to stage or re-stage relationships amongst relatives and friends. The doing of photography is supported by the material object of the photograph itself to produce what Richard Chalfen (1987) calls *'Snapshot Versions of Life'*. The holiday photograph as produced within a setting of intimacy is also largely consumed within that same setting. Even on occasions when the camera is temporarily loaned to a stranger to take a photograph in order to feature a full group or couple (with the favour often being reciprocated), a certain degree of trust is displayed and intimacy is the result (Sandle 2003). Indeed, the holiday photograph which features no people, only a building or panoramic vista, is as likely to be explained and discussed with reference to the intimate sharing of the view at the time the photograph is taken and/or later shown, in what Tim Edensor (1998, p. 126) refers to as 'communal witnessing'.

The 'local' is, of course a common subject of tourist photography and can be 'directed' to perform (often for a price), for the tourist. Staging a photograph which locates the tourist/stranger amongst local people is a way to contextualize and 'eternalize' a visit (Cohen, Nir and Almagor 1992). This is very much a marker of being outside of the everyday experience but it does entail what Erik Cohen et al. (1992) term 'photographer–photographee interaction', an ambiguous process of social exchange. Cohen et al. (p. 231) differentiate between contextualized and decontextualized photographs. The former, they argue, will have 'primarily a subjective meaning for the individual and serve as souvenirs'. This would include, in their words, a photograph that has been 'insidiously staged, such as snapshot of a busy street or market in a Third World country'. The latter, while also capable of acting as a souvenir, would 'possess a more general significance, as a document, even for those who did not share the experience of the visit'. This, in their view, would include the photographs taken by anthropologists and would be embedded in a wider and longer relationship with the locals.

While there are undoubted differences in the ways in which photography is performed and, differences between the motivations of its exponents, photography *is* a practice of identity construction ('othering') for both photographee and photographer. The tourist has clearly power in mediating local cultures through photographs, though this power is arguably exercised not in any knowingly exploitative way but as a function of normative human curiosity. Moreover, the local also has a capacity to appropriate and manipulate the tourist photographer in drawing their attention to specific sights and engaging in economic exchange. The tourist as photographer can also experience the 'reverse gaze' from the local, which puts the tourist into a position of emotional discomfort and reflexivity as the subject stares back (Gillespie 2006).

Materiality and Memory

Tourists, by definition, return home. While all life is a temporally bound condition, tourism is a particularly short one; a festive time that seems to allow people to recreate sense and significations associated with their lives (Graburn 1976). In the post-tourism context, the very personal moments of the holiday experience and the stories that emanate from these are ultimately transformed into more stable tourism memories and souvenirs. Stories are told to friends, work colleagues and relatives. Objects collected or acquired during the journey are shown around, sometimes given away as gifts, sometimes integrated to pre-existing personal souvenir collections or filing systems. In this sense the intangibility of memory and recollection is made tangible through the acquisition, presence and frequent interrogation of objects (Picard, forthcoming). The acquisition of photographs is of course almost exclusively de-limited to personal or family photographs providing us as Sontag (1977, p. 155) puts it, with a 'surrogate possession of a cherished person or thing, a possession which gives photographs some of the character of unique objects'.

In the act of recalling the holiday as a series of events and experiences of varying levels of significance, the photograph has become both a central prop and a prompt with its own materiality and life history. Photographs are printed, sometimes remaining in the developer's sleeves, arranged together with other photo-prints of previous events or holidays, sometimes put in elaborate albums with accompanying captions and descriptions. Holiday photographs can be copied and given as gifts, framed as large prints and put on the walls of living rooms. Increasingly they are put into virtual web-based photo albums made accessible to a public audience. All such actions indicate the strong social validatory role of holiday photographs in that they allow for the communication and projection of the self. The ritual of talking through a recent holiday with friends or family using holiday snaps (as prints, on a computer, and less frequently now as an illustrated talk using slides) as a narrative device, is illustrative of not only the process of validating the fact that one has visited a destination in the past but also the fact that one is participating in the social life of the present (Paster 1996).

Edward Bruner (2005) refers to the role of the tourist photograph as a mnemonic device for storytelling, but as he observes, the tales which are generated by the object are not simply a replay of events but 'embellish, privatize and transform the master narrative'. (p. 24) Bruner thus implicates the photograph in the production and circulation of the conceptual frames which shape our (personal) understandings of tourism (Bruner 1986). The public images of tourist sites and attractions, circulating in the world, and ostensibly for tourists – the standard, classical and 'traditional' pictures of tourist sites – are themselves made private by each tourist. In acts of both reflexive mimicry and oblivious innocence, the Pyramids at Giza, for instance, are taken into private ownership and imbued with particulars of momentary and intimate experience. The master narrative in almost a non-conscious way is thus perpetuated. At the same time the private life of the holiday photograph is shared. Still observable on the high streets of many towns and cities at the end of the summer holiday 'season', outside of the places where photographs are 'developed', or more commonly now, printed, is the almost immediate exhibiting of the holiday photographic record to friends and family.

More likely to take place around the digital camera as frames are re-played, stories of the holiday are told and diffuse through social networks as part of the very production of tourism.

While the journey to and through 'destinations' has allowed a temporary nullification of difference and separation, the act of leaving and returning back home seems to help in re-establishing these forms of separation. The collectively set frame of the journey, produced through various travel agents and operators, paced by the liturgy of travel representations, texts and forms of conduct has allowed tourists to bring the quasi-mythical realm of the destination alive, to create festive moments during which separations disappear (Caillois 1950), to connect to the distanced worlds of temporal, spatial and social Others and to recreate themselves as persons and social subjects. In the post-tourism context, the journey is coming to an end and its experiences are re-integrated into the context of quotidian life. Tourism photography and other forms of souvenir making seem to play an important role in relating the specific time-space of tourism to the quotidian life back home. Both seem capable to carry qualities associated with the destination into the home spaces of tourists. This leads back to the question of what precisely is captured in holiday photography?

People in all epochs and times have used specific objects or materials to encapsulate the power and mystical energies associated with places, people or presences and to transport them from one place to another. Specific objects have been used to 'capture' spirits and make, or help them travel to different places. Body parts, or objects in contact with body parts or bodies were believed to contain the spirit of the deceased, thus the preservation of bones, hair and finger nails were frequently used to capture the spirit of the departed. In some societies, the symbolic (and sometimes literal consumption) of the body was thought of as empowering consumers with spiritual power. The catechism of the Roman Catholic Church prescribes methods of reproducing relics that preserve their authenticity and efficiency, 'third class relics' which are created by touching items associated with events of the Christ's life, the physical remnants of a saint or an item that a saint wore (Sumption 1975). Religious pilgrims visiting religious sites or shrines of important relics usually acquire small objects, medals or images, or decant drops of the Holy water in plastic bottles, which they would eventually take home and integrate into their home space living room shrines or use for other spiritual-therapeutic purposes. Something similar seems to happen when tourists take photographs and display them.

From this perspective, photography, and tourism photography in particular, continue to be thought of, by largely common understanding, as technologies of encapsulating some form of 'authenticity'. From this perspective, photographic copies are far from being emptied of any sense or '"authentic" or "auratic" power' (Benjamin 1999), but on the contrary, enable their authors to transport such powers to their homes and circulate them among friends, relatives and wider audiences. In this sense, the power of the photograph is amplified by the very distance between the place of its capture and the tourists' home. Michael Jackson (1998) borrows

Figure 1.6   A forgotten performance, an 'anywhere' destination, with no reference point to time

Note: Outside of the context of family and friends, photographs would seem to have little in the way of public significance.

Mark Twain's aphorism that 'distance lends enchantment' to emphasize the metaphorical power of that which is normally beyond our reach; the more exotic or reified the destination and experience, the more power is imbued in the souvenir or photograph. Back within the normative frame of 'home' one only has to show and touch a photograph to invoke this magic-like power which has the capacity to generate memories, stories and 'uplifting experiences' (Edwards 1999).

The spontaneity of the holiday photograph, generally uncontaminated by technical over-indulgence, makes it one of the most honest of touristic objects. It seldom seeks to deceive the viewer. The producer of the holiday snap is usually also the main consumer. Seldom are tourist photographs burdened with recorded elaborations of the where, the when and the why of their existence. What narration exists around the photographic object also starts in a spontaneous and fleeting moment, and is easily lost. While it may prompt memory and storytelling, it is not the sole precondition for these. But while accepting the potential of the photograph to act as an object of memoriam and ignite a whole host of meanings, the fate of the holiday photograph is not always one of privilege or priority. Many (and probably the majority) of holiday photographs are destined to be unceremoniously dumped into drawers, and shoe boxes stored in attics. Unsorted and seldom looked at photographs can reside at the edge of obscurity for generations and are often only liberated through family deaths or moving house. The holiday photograph can lose its vitality over time; distance in the sense of passing years can also dull the notion of enchantment. The images captured are also forgotten, their symbolic status lost as memories fade and falter. Arguably, in the rapid expansion of the number and frequency of holidays taken, the significance of marking these as 'special' events through the taking of photographs has lessened in parallel with the loss of exclusiveness. The experiences of the holiday are, more than ever, capable of being reproduced by multiple visits to the same destination, in new destinations or indeed in 'any' destination. With the onset of the digital camera and the capacity to take multiple snapshots and edit/remove them later via the camera or the computer, meaning can swiftly drain from the material object of the photograph (Figure 1.6).

Michael Haldrup and Jonas Larsen (2006) make a persuasive case for researchers to go beyond the representational role of photography in tourism. In this sense they unwittingly appear to create a distinction between what we may term 'tourist photography' and the 'photography of tourism'. The former is grounded in the performances and socialities of individuals and groups of tourists replete with relationships shared between the materialities of photography, the intangibility of memory and, as they stress, the 'hybridity' of tourist–camera relations (Haldrup and Larsen 2006, p. 282). The latter are not contingent on any real sense of 'being there' and exist within the global, social sphere of representation, are seldom included in personal memory and are remote from the tourist's own creative practices, though as discussed earlier, their essentialism can be reflected in, and refracted by, the photographs that tourists take as part of the 'circle of representation'. These are the essential representations of holidays which are made visible, circulated widely and which are designed to elaborate what is worthy of viewing (Dann 1996). As William Cannon Hunter (2008, p. 357) puts it:

> Representations in the form of photographs have become fundamental to the very reality of tourism. They are foundational components of imagery and the mechanism of its discourses.

Elizabeth Edwards and Janice Hart (2004) remind us that the presentation of images in the nineteenth and early twentieth centuries, along with the materiality of their presentation, reflected the social saliency and quasi-religious status of the photograph. Furthermore, within the historical context, the circulation of tourist or travel photographs, as objects and souvenirs, was indeed significant in shaping our knowledge of the world, not explicitly through their documentary power but by their capacity to represent and create symbolic capital (Tagg 1993; Edwards 1996).

In the ideas around objects having 'social lives' (Appadurai 1986) and some manner of agency, the tourist photograph would seem to occupy a rather peripheral position. While capable of being re-contextualized and, in an artistic way, capable of being instrumental in eliciting an emotional response (Gell 1998), the tourist/family photograph needs human agency to provide it with significance (Steiner 2001). But this is, to marked extent, a significance that is normally elaborated within the private sphere, in the immediacy of the family and between generations, working with close memories, making memories tactile (Spence and Holland 1991; Edwards 1999), and circulating meanings amongst intimate networks. The voice of the object is seldom heard beyond the family album or its shoebox home. Outside of the social context of family and friends, photographs would seem to mean little, if they are displayed at all. What seems to be important in all of this are the social practices of tourist photography which, as a way of participating in/'producing' the tourism phenomenon, and as a collective referent of cultural/ inter-cultural engagement, are also entry points into uncovering the broader issues of sociality in tourism.

Understandings of the Tourism/Tourist Photography Relationship

The foregoing discussion merely serves to open further routes for interpretation and interrogation of the tourist, tourism, photography relationship. Any attempt at rigid categorization relating to the various approaches taken in the examination of tourism and photography would seem to be somewhat futile, such are the inter-sections and overlaps and overlays which occur. In this book we have invited authors to address their interests in tourism and photography from different historical, anthropological and sociological perspectives. In doing so, the common themes of power, play, performance, relationship with place and memory are all noticeable and are frequently approached with the context of the colonial. The primacy given to colonial relations as a framework for analysis merely mirrors the historical centrality of photography in constructing the colonial world (Ryan 1997; Hight and Sampson 2002; Hayes 2002; Johnston 2002; Thompson 2006). However, as a number of contributors address, what is significant are the ways in which historical patterns remain inscribed within the contemporary circulations of tourism and tourists.

In Chapter 2, Matthew Martinez and Patricia Albers offer a diachronic study of the visual techniques, ideologies and power relationships underlying the touristic enchantment, imaging and imagining of Pueblo people in Northern New Mexico. Nineteenth century tourism photography is seen here as a powerful means of generating, diffusing and controlling a conventionalized representation of Pueblo people and their place within the wider social worlds of North American society. Through the subtle use of realist photographic style, romanticizing voice-overs and the choice of pastoral photographic backdrop themes, the Pueblo people subsequently become stereotyped as a bread-baking and ceremonial dance-performing people, living within the wider realm of an a-historical, mythical New Mexico. Through the development of itineraries and sites, tourism consequently reaffirmed and further homogenized the stereotypes underlying these essentialist images. While the tropes of the romantic wilderness, and of a Western frontier continued to underscore tourism representations of the Pueblo people during the twentieth and into the twenty first century, a shift in style and authorship takes place. With the progressive social and political emancipation of the Pueblo people, this is translated by the use of more 'ennobling' photographic styles and, more importantly, the taking of control by the Pueblos over the production of their visual representations. By restricting tourist access to certain ceremonies and social spaces, and by producing and authoring their own guidebooks, the Pueblos increasingly assert control over the way they are represented within the tourism realm. At the same time, they actively promote alternative visions of themselves as people participating in contemporary history, conscious of the future. These visions often engage with the pastoral and frontier tropes underlying the former colonial representations of the Pueblo but, they are integrated, sometimes playfully, sometimes seriously, in an auto-ethnographic way, into the current self-fashioning of the Pueblo, out of the colonial context and towards liberation.

Vassiliki Lalioti, in Chapter 3 analyses how the Greek National Tourism Organization uses visual representations of 'Greekness' to define and position Greece as a tourism destination within an international, and primarily Western European context. Through a historical and ethnographic study of ancient Greek theatres as 'visual images', Lalioti reveals the genealogy of the 'semiosphere' in which Greece and its capital, Athens, evolved and were touristically enchanted as the birthplace of secular European civilization. In this context, controlling the image of Greece's past, the integrity of the material relics embodying this past, and the touristic access to these relics have become major political and cultural issues within Western Europe. From there, Lalioti suggests that we see modern 'Greekness', as being the 'perpetual cultural matrix of Europe', as fabricated through the political and touristic dialectic between Western European centres and the Greek periphery. This fabrication is visually achieved by essentializing on two tropes: the idea of 'eternal sunshine' and the de-historicized realm of 'ancient culture'. In this sense, contemporary tourism advertisements typically include images of antiquities in full sunlight and thus continue to reproduce these tropes. Being the living ancestor of Western civilization within an enchanted land in the periphery of Europe has long become a generic marker for Greece as a tourism destination. In a more ambiguous way, this role has also been embraced within the making of Greek national identity, differentiating Greece within the wider political and social context of Western Europe.

Chapter 4 deals with persistent and pervasive colonial representations and their global cultural context. Brian Cohen and Ilyssa Manspeizer study the dilemma faced by Western agencies working in the fields of relief and development; that of picturing 'Africa'. In particular, they focus on the inability of these agencies to overcome specific Western cultural frameworks that have historically determined how Africa should be seen and 'known'. From this point, even 'well-intentioned' photographs used in fundraising campaigns tend to perpetuate the colonial vision of a child-like and, not-quite-human, Africa, a vulnerable Africa 'crying for help'. The effect of these images is obtained by projecting Western literary tropes into African contexts, to generate meaningful visual 'stories'. Accordingly, Western media usually construct Africa either through an Afro-romantic trope of untamed nature and a-historic existence, or through an Afro-pessimist trope of chaos, drought, famine and war. These visual 'stories' are often further enhanced through references to the great themes of Western art. While Westerners in such photos regularly appear as heroic 'icons of salvation', Africans remain 'often faceless figures who serve as the objects of Western beneficence'. For Cohen and Manspeizer, the idea of Africa as a 'doomed' continent, that served colonialism so well, today endures in popular culture and especially in tourism and philanthropic photography. On a global scale, Africans continue to be attributed the role of the 'other' as not quite 'us' – validating further injustices against them.

Stan Frankland, in Chapter 5, studies the tourist consumption of Pygmies as an ethnic tourism product and the way in which photography is implicated in this process. For Frankland, the touristic value of the Pygmy 'brand name' resides precisely in their social and ethnic 'otherness' satisfying the touristic search for 'untouched' environments and populations. In this context, the trope of loss, typically embodied in the idea that 'any Eden found must soon be lost' remains a structuring motif in Western thinking. From there, tourism appears to keep alive earlier forms of colonial nostalgia and the idealization of 'savage' peoples living somehow suspended in a perpetual present, threatened by the course of time. This trope of loss projected into the idealized existence of the Pygmy and other 'savage peoples' becomes the essence of the contemporary ethnic tourism product (Albers and James, 1983). Sustaining an image of the 'doomed' consequently becomes the single most important issue in maintaining the tourism semiosphere, both for the Pygmy and for Western tourists. To describe the circularity of the tourism consumption and production processes, Frankland uses the Lévi-Straussian metaphor of symbolic cannibalism. As with food, the digestive process of consumption leads to a homogenization of the consumed 'other' and consequently to the nullification of difference. To avoid this and re-establish the fundamental principal of difference defining the Pygmy within the tourist semiosphere, tourists symbolically regurgitate the consumed 'other'. Vomit, Frankland writes, has a regenerative quality – it does not mark the end of the process of consumption, instead it signifies the beginning of a new phase. The never-ending circularity of consumption and regurgitation hence allows to maintain the myth of the Pygmy, but also the people themselves, in a condition of extreme difference. Once established as an iconic figure of the colonial and tourism semiospheres, this circularity enables the Pygmy to progressively integrate different discursive functions and tropes: 'from savage exemplars into eco-saviours, from prototypes of the primitive mind into protectors of a forgotten wisdom, and from racist curiosity into relative wonder'.

Also within the seemingly constant context of colonialism, Elvi Whittaker, in Chapter 6 analyses how tourism photography within the colonial realms of the late nineteenth and early twentieth centuries established a pervasive visual discourse about race and human difference. In these contexts, photographic technology was considered as truth-recording machinery, able to objectify images of the human condition and its presumed discontinuities. Photography consequently became a powerful tool in the service of the colonial and scientific race knowledge doctrines. Used to document racial stereotypes, it aesthetically helped to legitimize the Western discourse on the inequality of human races and the colonial project of submitting non-Western people, especially the 'doomed races', to the 'civilizing' and assimilating mission of the colonial empires. In this sense, Whittaker writes, photography was instrumental in supporting every notion of Western superiority. Tourism photography was complicit in the dissemination of racial stereotypes and the wide diffusion of the colonial 'race knowledge'. Through artistic techniques of formal reduction, simplification, and extraction from cultural references, the photographed subjects were rendered into different types of conventionalised compositions. Indigenous members of presupposed 'doomed' races and cultures, but also the lower European classes, were usually pictured in 'lowly' ways, with photographs usually being built around the visual tropes of fixity, decay, sexual availability, and social inferiority. At the same time, the elites and ruling classes of indigenous peoples were usually pictured in the noble pose used to frame European elites. Consequently, Whittaker argues, the touristic techniques of photographing race transposed the aesthetic codes of picturing members of different European classes and socio-economic status into the colonial realms. The underlying discourse on the inequality between social classes is projected into this realm, as a classifying system to discriminate higher and lower, 'noble' and 'doomed', races/ classes/cultures and legitimize the global colonial semiosphere.

In Chapter 7, Teresa E.P. Delfin studies how the visual rhetoric of tourism advertisements influences the imaginings and symbolic constructions of tourism reality. She argues that the visual rhetoric of tourism advertising goes beyond the colonial trope of empty land, by marketing a trope of palimpsests, with layers of meaning to negotiate, and contribute to. From there, tourist sites, despite any officially sanctioned meanings or marketing images, become 'virtual halls of mirrors' enabling tourists to make them conform to their fantasies. One of the more dominant 'fantasies' played out within the tourism realm is the idea of 'travelling through the past'. In many ways, the idea of the past has become a 'site of escape'. Imaginations and idealizations of how life was in the past are consequently projected into foreign landscapes and people. Yet the visual rhetoric of advertising images seems very clear about the phantasmagorical dimension of such a journey and the power relations played out between the traveller and the travelled. The 'past' comes into life as an act of post-touristic performance. This is not about the fruitless search of an 'authentic' past, but about the act of performing a fantasy past. In this play of projections, the tourist is as much an actor, as are the visited landscapes and people. Because of the fantastic nature of tourism, Pellinen-Chavez writes, tourist imaginaries do have material consequences on the spaces tourists occupy and on the people with whom they may or may not come into contact, but with whom there is nonetheless a connection based on the economic structures of tourism. Destinations consequently are dependant, to a certain degree, on creating or maintaining conditions and settings that allow the mirroring of the tourist fantasy. Developing countries or regions acting as 'living pasts' within the tourism realm are, in particular, confronted with the problem of maintaining a 'pre-modern' setting mirroring tourism fantasy while at the same time offering modern accommodation, transport, safety and health infrastructures.

Janet Hoskins explores in Chapter 8, the colonial stereotype of the native who 'fears that the camera will steal his soul' and the alternative phantasm of 'global vampirism', both corresponding to a discursive formation in colonial thought. Colonial writings of the nineteenth century recurrently document stories about 'local superstitions' regarding the power of a photo camera to absorb into itself some of the vitality of photographed people or sacred objects. More recent ethnographic writings from different locations in the world report local fears of having their blood or fat sucked out by photo cameras. Through her own fieldwork in Indonesia, Hoskins follows local stories of blood sucking photo cameras, with children being suspended upside-down inside; stories she explains by the upside-down images in early photographic boxes and the visual analogy to Muslim modes of animal sacrifice and butchering. While in all of these stories, some vital fluids – blood and fat – is believed to being extracted from the body of the photographed, Hoskins asserts that the theory of a soul-stealing superstition as reported in colonial writings and perpetuated today through contemporary tourism practice, is a romanticizing Western fantasy. Hoskins argues that beyond these predatory fantasies and phantasms, the photo camera works as a vehicle for re-imagining reality within post-colonial contexts. As a technology of enchantment, photography enables former colonial people to invert the circumstances of the 'imperial spectacle' and hence becomes a tool to transform local visions and stories. From there, Hoskins concludes, the contemporary formulation of accounts of enchanted cameras can be seen as a form of critical questioning of visual inequity rather than the local worshipping of Western technological advances.

In Chapter 9 Andy Letcher, Jenny Blain and Robert Wallis analyse the contested interpretations and uses of stone circles in Stonehenge and Avebury, in England. Through a critical discourse analysis of pictorial representations by conservation agencies, the authors reveal how these stone circles are being produced and consumed as images of prehistoric sites. Through a combination of distanced panoramic shots, emptied of any human presence, and close-ups of the actual stones, the sites begin to embody an idea of the past as a closed, unreachable and permanent thing. In this conception, any change or alteration of the stones is seen as 'degradation', alienating the site from its visual ideal and its underlying conservationist ethos. The romantic trope underlying the officially sanctioned representation of the site, stressing distance and temporal fixity, is further amplified by cosmetic landscape surgery. Non-prehistoric elements – a visitor centre and streets – are physically relocated and thus removed from the vision. Yet, the authors stress, the stone circles have never been a singular thing, but a palimpsest, a continually changing composite. In the contemporary context, the conservationist ideal of the site as a dramatic landscape in which humans no longer play a part is increasingly contested, sometimes violently. In a time where the distant past seems to matter to people, pre-historic sites increasingly attract a heterogeneous crowd of tourists, new age pilgrims, earth mystics, pagans, and festival-goers. Letcher, Blain and Wallis conclude that the resulting conflict regarding access rights and ownership reveals a deeper conflict between knowledge and representation in which photography is complicit.

Photographs are a central instrument in the 'making' of tourist sites/sights. In Chapter 10, Celmara Pocock studies how photography has created the Great Barrier Reef in Australia as an object of tourism consumption. She shows how the multiple fragments of the reef could only be seen as a whole through the visual techniques of cartography and later aerial and satellite photography. Furthermore, this distant vision usually selects frames that match a generic reef imagery, modelled on the visual ideal of Pacific coral cays. Through the development of helicopter and plane excursions above the reefs, this vision has been made touristically accessible. The consumption of the distant vision of the reef is usually combined with a close-up vision, achieved usually through bodily immersion and the use of goggles. While scientific and naturalist expeditions produced watercolours and coloured black and white photographs of corals since the late nineteenth century, tourism underwater photography and scuba dive equipment are technologies that emerged during the 1960s and became popular among tourists only during the 1990s. For Pocock, photographic technologies of framing and timing have transformed the reef into a hyper-real space. The camera functions as a tool to capture the aesthetic emotion of seeing the 'indescribable' and to collect and catalogue what is 'beyond human imagination'. In this sense, photography replaces earlier activities of collecting, with the photography maintaining the 'essence' of the copy. The importance of images, writes Pocock, relates to the act of taking the image and being part of it.

Joyce Hsiu-yen Yeh, in Chapter 11, analyses tourism photography as a social act and a performance enabling the creation and mediation of sociality in both tourism and post-tourism contexts. Her study focuses on Taiwanese students she accompanied to England. She suggests that, for these students, photography is a way to record what they see, but also a way to construct and reaffirm social relationships within the social realms of both the actual tour and the post-tour home environment. In this interrelated home-abroad context, photography becomes a tool to capture and subsequently mediate the specific identity of 'being a tourist', as somehow distinct from daily practices. In this sense, capturing the tourist gaze as some sort of conventionalized or public-culture-sanctioned vision of tourist sites is part of this tourism identity building. It simultaneously allows transgression and plays with collective conventions, hence empowering the tourist to produce and mobilize culture, and allow a 'new' vision of culture to emerge. Photography, and other self-made souvenirs, are used to mediate social interactions, both in the tourism setting and later, back home. The photographic act hence becomes part of the wider tourism performance. The camera signifies the tourist within this act and the spaces in which it takes place. Within the social realm of the tour, it permits the creation of familiarity and a certain sense of shared community among tourists, but also engages with the visited strangers. Cameras contribute to the creation of a connection between self and others outside the group, Yeh writes, it is the camera that generates the interaction between tourists and strangers. In post-tourism contexts, pictures mediate the experience and become the physical support for stories. Furthermore, as encapsulations of memory, photographs – by looking at them or touching them – are used by some of the Taiwanese students to 'lift their spirit' in situations of boredom or dissatisfaction.

In Chapter 12 Elisabeth Brandin critically explores a method of interviewing tourists based upon their own photographs in order to study meanings of touristic practice as a form of embodied experience. Her experimental study focused on canoe tourists in Denmark to whom she handed out single-use cameras. Prints were developed after the trip and used during the interviews. The idea to develop such a method based on 'visual evidence' grew out of the challenge to address abstract emotional topics related to tourism. The photographs taken by the subjects of this study not only served as a narrative baseline to build the story of the day, but also prompted memories of experiences not captured in the pictures. In many ways, the pictures became vehicles to 'show' others experiences 'hard to describe with words'. In this sense, tourists use photographs as a tool to communicate highly embodied tourism practices. What Brandin demonstrates is how tourist photographs combined with interviews with their 'authors' can give access to tourism as a signifying practice.

Rebekah Sobel in Chapter 13 studies the progressive development of post-tourism narratives following forms of politically instrumentalized tourism to Israel. At the end of the 1990s, Birthright Israel, a Jewish philanthropic organisation sponsored free trips to Israel for more than 10,000 Jewish students from all over the world. According to Sobel, the aim of this was to challenge the growing assimilation of Jewish youths into their adopted countries' ways, of learning to strengthen their connection to Israel and to support the continuity of Judaism in the diaspora. Sobel follows a group of Georgetown University students during and after their trip to Israel. Looking at the photographs produced during this trip, she writes, is a tangible means to look at the connections between Israel and diasporic Jewish communities. It allows the study of how students connect to ideas of cultural nationalism, transnational identity and tourism, and evaluates their connection to these ideas over time. Talking about photographs in a non-focused way revealed itself to be a good research method. In many cases, one image triggered a story that led to another image, conveying another anecdote, hence inducing a communication cycle that reveals personal sentiments in an inductive manner. For Sobel, photographs become mediators for social and emotional links to a place. In this sense, during the trip, it was of 'incredible' importance for the students to place themselves in the photos. At the same time, the actual backdrop of the photographed site, seemed less important. Most photographs were taken not of the most anticipated or favourite sites, but of unexpected places, encounters or social activities. In the post-tourism context, photographs often embodied particular situations or happenings that could stand for a wider emotional context. Yet, over time, the descriptions of photographs and the stories or emotions they entailed, changed. Six months after the trip, Sobel writes, being Jewish and Israel are no longer separate concepts for the students. Both are now intertwined, part of the larger ideas of Jewish identity and community.

Arguably, one of the most influential concepts in the study of tourism, that of the 'tourist gaze', as used by John Urry (1990, 2002) is examined in Chapter 14 by Marie-Françoise Lanfant. Lanfant interrogates Urry's use of the term 'gaze' and reflects on it as being a departure point for a reflection on the complexity of what is shown, what is 'given to see' and what is actually seen in the field of tourism. Through an exploration of the concepts of imperception and scopic structure, she resituates the theory of the gaze in the French intellectual milieus of the post-war period which were heavily influenced by Husserl's writing about phenomenology. Refuting Bishop Berkeley's idea that the perception of the world leads to its essential truth, Lanfant sponsors an approach that goes beyond the principles of scopic vision and, the epistemology of a science of the gaze; an approach that firmly re-establishes the observing subject within the frame of observation and the exchange of gazes which this entails.

What all of these chapters reveal is the multi-layered and intersecting nature of tourist photographs and tourism photography; expansive categories which are at once separate and yet inter-related. The former is a particularly problematic category which oscillates between the representational 'surface' power of the photograph and its role in shaping visions of the world, to the processes and performances which attend the doing of photography. There is no pure approach to the interrogating of this category. It is defined by its messiness and the overlaps it shares with other substantive realms such as family and social life, with art and aesthetics, with imagined geographies, constructed and de-constructed identities and the implied complexities of power relations. Tourist photography is foremost photography in 'other' places and shaped by the ways in which tourists behave in such places. Patterns of ordinary life and normative social/family relations are not left at home but brought into the tourist space. They are played out in the being and doing of tourism, the performances of 'taking' of photographs, the using of cameras and, attendant playing, pointing and posing. Tourist photographs as objects, memory prompts and items of historical record are brought back into the rhythms of daily social life and are re-worked in new contexts and relationships.

Similarly the category of tourism photography, relating to the ways in which the producers, developers and promoters of tourism mobilize images, is similarly complex, overlapping as it does with a myriad of forms and formats through which places and peoples are represented. Boundaries are blurred between photographs of destinations and attractions used specifically to draw in tourists, between professional 'travel' photography, and between photographs used in wider advertising and for educational purposes. Photographs in this sense have fluid time frames; pictures of the Taj Mahal could have been taken today or one hundred years ago. In the context of colonial photography the legacy of inscription can be powerful and long lasting, shaping tourist imaginations many decades later. Photography as used in the provision and promotion of tourism frequently exists apparently outside of ownership, or at least away from the locatedness of time and place inscriptions and commentaries given to family holiday snaps. It flows easily across media; from the cover of a holiday brochure, to web-site, to a vast advertising hoarding, to a foreign magazine, to an extract in a DVD. It is defined by a ubiquity of being seen, what Baudrillard (1998) terms an 'ecstasy of communication', to the point where he argues that 'scene' becomes 'obscene' through its very 'transparence and immediate visibility' (p. 150).

What is clear from our own attempts to scope the multiple relationships between tourists, tourism and photography and, those of our contributors, is that we need to embrace an approach which can bring different disciplinary lenses to bear and a range of methodological approaches. We need to better understand the linkages and relations between tourist photography as practiced and performed at the individual and intimate familial level, and the uses, influences and mobilities of photography in terms of global tourism; linking agency and structure, local practices and global impacts. Moreover, we need to increasingly explore the many overlapping contexts, spaces and practices of tourist photography with photography in general so that we can better understand how tourists frame the world.

References

Adorno, T.W. and Horkheimer, M. (1997), *Dialectic of Enlightenment*, Verso: London.

Adler, J. (1998), 'Origins of Sightseeing', in: Williams, C.T. (ed.), *Travel Culture: Essays on What Makes Us Go*, Westport, CT: Praeger, pp. 3-23.

Albers, P.C. and James, W.R. (1983), 'Tourism and the Changing Photographic Image of the Great Lakes Indians', *Annals of Tourism Research* 10, pp. 123-148.

Albers, P.C. and James, W.R. (1988), 'Travel Photography: A Methodological Approach', *Annals of Tourism Research* 15, pp. 134-158.

Appadurai, A. (1997), 'The Colonial Backdrop', *Afterimage*, 24(5), pp. 4-7.

Bærenholdt, J.O., Haldrup, M., Larsen, J. and Urry, J. (2004), *Performing Tourist Places*, Aldershot: Ashgate.

Barthes, R. (1993), *Camera Lucida*, London: Vintage.

Baudrillard, J. (1997), 'The Ecstasy of Photography: Jean Baudrillard Interviewed by Nicholas Zurbrugg', in: Zurbrugg, N. (ed.), *Jean Baudrillard: Art and Artefact*, London: Sage Publications, pp. 32-42.

Baudrillard, J. (1998), 'The Ecstasy of Communication', in: Foster, H. (ed.), *The Anti-Aesthetic: Essays on Postmodern Culture*, New York, NY: The New Press, pp. 145-154.

Benjamin W. (1999), 'The Work of Art in the Age of Mechanical Reproduction', in: *Illuminations*, London: Pimlico, pp. 211-244.

Benjamin W. (2006), 'A Small History of Photography', in: *One Way Street – and Other Writings*, London: Verso, pp. 240-257.

Berger, J. (1972), *Ways of Seeing*, London: BBC/Harmondsworth: Penguin.

Berger, J. (1980), *About Looking*, New York, NY: Pantheon.

Boorstin, D.J. (1985), *The Discoverers*, New York, NY: Vintage Books.

Bourdieu, P. (1990), *Photography: A Middle Brow Art*, London: Polity Press.

Bruner, E.M. (2005), *Culture on Tour: Ethnographies of Travel*, Chicago, IL: University of Chicago Press.

Caillois R. (1950), *L'Homme et le sacre*, Paris: Gallimard.

Chalfen, R. (1987), *Snapshot Versions of Life*, Bowling Green, OH: Bowling Green State University Popular Press.

Chambers, D. (2003), 'Family as Place: Family Photograph Albums and the Domestication of Public and Private Space', in: Schwartz, J. and Ryan, J. (eds), *Picturing Place: Photography and the Geographical Imagination*, London: I.B. Tauris, pp. 96-114.

Cohen, C.B. (1995), 'Marketing Paradise, Marketing Nation', *Annals of Tourism Research*, 22, pp. 404-21.

Cohen, E., Nir, Y. and Almagor, U. (1993), 'Stranger-Local Interaction in Photography', *Annals of Tourism Research*, 19, pp. 213-233.

Crang, M. (1997), 'Picturing Practices: Research through the Tourist Gaze', *Progress in Human Geography*, 21(3), pp. 359-373.

Crouch, D. (2000), 'Places Around Us: Embodied Lay Geographies in Leisure and Tourism', *Leisure Studies*, 19, pp. 63-76.

Crouch, D. (2002), 'Encountering Space', in: Crang, M. and Coleman, S. (eds), *Tourism: Between Place and Performance*, London: Berghahn Books: pp. 207-218.

Crouch, D. and Lübbren N. (eds) (2003), *Visual Culture and Tourism*, New York, NY and Oxford: Berg.

Crouch, D., Jackson, R., and Thompson, F. (eds) (2005), *The Media and the Tourist Imagination: Converging Cultures*, London: Routledge.

Dann, G. (1996), 'The People of Tourist Brochures', in Selwyn, T. (ed.), *The Tourist Image: Myths and Myth Making in Tourism*, London: John Wiley & Sons Ltd, pp. 61-81.

Duro, P. (1996), *The Rhetoric of the Frame: Essays on the Boundaries of the Artwork*, Cambridge Studies in New Art History and Criticism, Cambridge: Cambridge University Press.

Edensor, T. (1998), *Tourists at the Taj*, London: Routledge.

Edensor, T. (2005), 'Waste Matter – The Debris of Industrial Ruins and the Disordering of the Material World', *Journal of Material Culture*, 10(3), pp. 311-332.

Edwards, E. (1996), 'Postcards: Greetings from Another World', in Selwyn, T. (ed.) *The Tourist Image: Myth and Myth-Making in Tourism*, Chichester: John Wiley & Sons, pp. 197-221.

Edwards, E. (1999), 'Photographs as Objects of Memory', in Kwint, M., Breward, C. and Aynsley, J. (eds), *Material Memories: Design and Evocation,* London: Berg Publishers Ltd, pp. 221-236.

Edwards, E. and Hart, J. (eds) (2004), *Photographs Objects Histories: On the Materiality of Images*, Routledge: London.

Flusser, V. (2007), *Towards a Philosophy of Photography*, London: Reaktion Books.

Garlick, S. (2002), 'Revealing the Unseen: Tourism, Art and Photography', *Cultural Studies*, 16(2), pp. 289-305

Gautrand, J.C. (1998), 'The Traveler's Paraphernalia', in: Frizot, M. (ed.), *The New History of Photography*, Köln: Könemann, p. 158.

Gell, A. (1998), *Art and Agency: An Anthropological Theory*, Oxford: Clarendon Press.

Gell A. (2005), 'The Technology of Enchantment and the Enchantment of Technology', in: Coote, J. and Shelton, A. (eds), *Anthropology: Art and Aesthetics*. Oxford: Oxford University Press, pp. 40-63.

Gillespie, A. (2006). 'Tourist Photography and the Reverse Gaze', *Ethos*, 34(3), pp. 343-366.

Gombrich, E.H. (1972), *Art and Illusion*, Princeton, NJ: Princeton University Press.

Goss, J. (1993), 'Placing the Market and Marketing Place: Tourist Advertising of the Hawaiian Islands, 1972-92', *Environment and Planning, Society and Space*, 11, pp. 663-88.

Graburn, N. (1976), 'Tourism: The Sacred Journey', in: Smith, V. (ed.), *Hosts and Guests*. Philadelphia, PA: University of Phildelphia Press, pp. 22-36.

Gregory, D. (1995), 'Imaginative Geographies', *Progress in Human Geography,* 19, pp. 447-485.

Gregory, D. (2003), 'Emperors of the Gaze: Photographic Practices and Productions of Space in Egypt, 1839-1914', in: Ryan, J. and Schwartz, J. (eds), *Picturing Place: Photography and the Geographical Imagination*, London: I.B. Tauris, pp. 195-225.

Haldrup, M. and Larsen, J. (2003), 'The Family Gaze', *Tourist Studies*, 3(1), pp. 23–46.

Haldrup, M. and Larsen, J. (2006), 'Material Cultures of Tourism', *Leisure Studies,* 25(3), pp. 275-289.

Hallman, B.C., Mary, S. and Benbow, P. (2007), 'Family Leisure, Family Photography and Zoos: Exploring the Emotional Geographies of Families', *Social and Cultural Geography*, 8(6), pp. 871-888.

Harrison, J. (2003), *Being a Tourist: Finding Meaning in Pleasure Travel*, Vancouver: University of British Columbia Press.

Hayes, M. (2002), 'Photography and the Emergence of the Pacific Cruise: Rethinking the Representational Crisis in Colonial Photography', in: Hight, E.M. and Sampson, G.D. (eds) (2002), *Colonialist Photography: Imag(in)ing Race and Place*, London: Routledge, pp. 172-187.

Hight, E.M. and Sampson, G.D. (eds) (2002), *Colonialist Photography: Imag(in)ing Race and Place*, London: Routledge.

Hirsch, M. (1997), *Family Frames: Photography, Narrative and Postmemory*, Cambridge, MA: Harvard University Press.

Hirsch, M. (1999), *The Familial Gaze*, Hanover, NH: University Press of New England.

Holland, P. (1991), 'Introduction: History, Memory and the Family Album', in: Spence, J. and Holland, P. (eds), *Family Snaps: The Meanings of Domestic Photography*, London: Virago, pp. 2–12.

Holland, P. (2000), 'Personal Photography and Popular Photography', in: L. Wells (ed.), *Photography:A Critical Introduction*, London: Routledge, pp. 117-162.

Hom Cary S. (2004), 'The Tourist Moment', *Annals of Tourism Research*, 31(1), pp. 61-77.

Hunter, W.C. (2008), 'A Typology of Photographic Representations for Tourism: Depictions of Groomed Spaces', *Tourism Management*, 29(2), pp. 354-365.

Hutnyk, J. (1996), *The Rumour of Calcutta: Tourism, Charity and the Poverty of Representation*, London: Zed Books.

I'Anson, R. (2000), *Travel Photography: A Guide to Taking Better Pictures*, London: Lonely Planet Publications.

Jackson, M. (1998), *Minima Ethnographica: Intersubjectivity and the Anthropological Project*, University of Chicago Press: Chicago

Jager, J. (2003), 'Picturing nations: landscape photography and national identity in Britain and Germany in the mid-nineteenth century', in: Ryan, J. and Schwartz, J. (eds), *Picturing Place: Photography and the Geographical Imagination*, London: I.B. Tauris, pp. 117-140.

Jenkins, O.H. (2003), 'Photography and Travel Brochures: The Circle of Representation', *Tourism Geographies*, 5(3), pp. 305-328.

Johnston. P. (2002), 'Advertising Paradise: Hawai'i in Art, Anthropology and Commercial Photography', in: Hight, E.M. and Sampson, G.D. (eds) (2002), *Colonialist Photography: Imag(in)ing Race and Place*, Routledge: London, pp. 189-225.

Kotchemidova, C. (2005), 'Why We Say "Cheese": Producing the Smile in Snapshot Photography', *Critical Studies in Media Communication*, 22(1), March 2005, pp. 2-25.

Larsen, J. (2005), 'Families Seen Photographing: The Performativity of Tourist Photography', *Space and Culture*, 8(4), pp. 416-434.

Larsen, J. (2008), 'Practices and Flows of Digital Photography: An Ethnographic Framework', *Mobilities*, 3(1), pp. 141-160.

Latour, B. (1991), 'Technology is Society Made Durable', in: J. Law (ed.), *A Sociology of Monsters: Essays on Power, Technology and Domination*, London: Routledge, pp. 103-131.

Lippard, L.R. (1997), *The Lure of the Local: Sense of Place in a Multi-centered Society*, New York, NY: New Press.

Löfgren, O. (1999), *On Holiday: a History of Vacationing*, Berkeley, CA: University of California Press.

Lutz, C. and J. Collins (1993), *Reading National Geographic*, Chicago, IL: University of Chicago Press.

MacCannell D. (1976), *The Tourist: A New Theory of the Leisure Class*, New York: Schocken.

Markwick, M. (2001), 'Postcards from Malta: Image, Consumption, Context', *Annals of Tourism Research*, 1(2), pp. 417–438.

Markwell, K. (1997), Dimensions of Photography in a Nature-Based Tour, *Annals of Tourism Research*, 24(1), pp. 131–155.

Micklewright, N. (2003), *A Victorian Traveler in the Middle East: The Photography and Travel Writing of Annie Lady Brassey*, Aldershot: Ashgate.

Mirzoeff, N. (1998), 'What is Visual Culture?', in: Mirzoeff, N. (ed.), *The Visual Culture Reader*, London: Routledge, pp. 3-13.

Osborne, P. (2000), *Travelling Light: Photography, Travel and Visual Culture*, Manchester: Manchester University Press.

Parr, M. (1995), *Small World*, London: Dewi Lewis Publishing.

Paster, J. (1996), *Snapshot Magic: Ritual, Realism and Recall*, Detroit, MI: UMC.

Picard, D. (forthcoming), *Tourism, Magic and Modernity: Cultivating the Human Garden*, Oxford: Berghahn

Ryan, J.R. (1997), *Picturing Empire; Photography and the Visualisation of the British Empire*, London: Reaktion Books and Chicago, IL: University of Chicago Press.

Sandle, D. (2003), 'Joe's Bar, Douglas Isle of Man: Photographic Representations of Holiday-makers in the 1950s', in: Crouch, D. and Lübbren N. (eds), *Visual Culture and Tourism*, New York, NY and Oxford: Berg.

Schechner, R. (1988), *Performance Theory*, London: Routledge.

Schwartz, J.M. (1996), 'The Geography Lesson: Photographs and the Construction of Imaginative Geographies', *Journal of Historical Geography*, 22(1), pp. 16-45.

Schwartz, J. and Ryan, J.R. (eds) (2003), *Picturing Place: Photography and Imaginative Geographies*, London: I.B.Tauris.

Shields, R. (1991), *Places on the Margins: Alternative Geographies of Modernity*, London: Routledge.

Snow, R. (2008), 'Tourism and American Identity: Kodak's Conspicuous Consumers Abroad', *Journal of American Culture*, 3(1), pp. 7-19.

Sontag, S. (1977), *On Photography*, London: Penguin Books.

Spence, J. and P. Holland (1991) (eds), *Family Snaps: The Meanings of Domestic Photography*, London: Virago.

Strain, E. (2003), *Public Places, Private Journeys: Ethnography, Entertainment, and the Tourist Gaze*, New Brunswick, NJ: Rutgers University Press.

Steiner, C. (2001), 'Rights of Passage: On the Liminal Identity of Art in the Border Zone', in: Myers F. (ed.), *The Empire of Things: Regimes of Value and Material Culture*, Santa Fe, NM: School of American Research Press, pp. 207-231.

Steward, S. (1993), *On Longing. Narratives of the Miniature, the Gigantic, the Souvenir, the Collection*, Durham and London: Duke University Press.

Sumption, J. (1975), *Pilgrimage: An Image of Medieval Religion*, London: Rowman and Littlefield.

Tagg, J. (1993), *Burden of Representation: Essays on Photographies and Histories*, Minneapolis, MN: University of Minnesota Press.

Taussig, M. (1987), *Shamanism, Colonialism and the Wild Man*, Chicago, IL University of Chicago Press.

Terdiman, R. (1993), *Present Past: Modernity and the Memory Crisis*, Ithaca, NY: Cornell University Press.

Thomas, K. (1975), 'The Tin-Box Photos', *Saudi AMRACO World Magazine*, 26(5), pp. 26-32.

Thompson, K. (2006), *An Eye for the Tropics: Tourism, Photography, and Framing the Caribbean Picturesque*, Durham, NC: Duke University Press

Tomlinson, J. (2007), *The Culture of Speed: The Coming of Immediacy*, London: Sage.

Turner, V. (1990), 'Are There Universals of Performance in Myth, Ritual and Drama?', in: Schechner, R. and Appel, W. (eds), *By Means of Performance: Intercultural Studies of Theatre and Ritual*, New York, NY: Cambridge University Press, pp. 8-18.

Urry, J. (1990), *The Tourist Gaze*, London: Sage.

Urry, J. (1995), *Consuming Places*, London: Routledge.

Urry, J. (2002), 'Globalising the Gaze', new chapter in second edition of *The Tourist Gaze*, London: Sage.

Urry, J. (2003), *Global Complexity*, Cambridge: Polity.

Wade, J, (1979), *A Short History of the Camera*, Watford: Fountain Press.

West, N. (2000), *Kodak and the Lens of Nostalgia*, Charlottesville, VA: University of Virginia Press.

Wilson, D. (1985), *Francis Frith's Travels: A Photographic Journey Through Victorian Britain*, London: Dent.

Willsberger, J. (1977), *The History of Photography: Cameras, Pictures, Photographers*, New York, NY: Doubleday.

<www.kodak.com>, accessed 14th July 2008.

Chapter 2
Imaging and Imagining Pueblo People in Northern New Mexico Tourism

Matthew J. Martinez and Patricia C. Albers

Introduction

Ever since the late nineteenth century, New Mexico has been portrayed as an *enchanted* and *enduring* land, a magical and ancient place where extraordinary forms of experience persist and where *exotic* cultures perpetuate themselves in spite of history (Bennett 1934; Spencer 1934; Anonymous, c. 1950s; Pomeroy 1957; Thomas 1978; Weigle 1989; Albers 1995; Dilworth 1996). The legendary imagining of New Mexico is closely tied to its identity as an internationally unique travel destination, and today, it continues to draw heavily on its indigenous peoples, especially those from the eight northern Pueblo communities of the Tiwa-speakers at Taos and Picuris, and the Tewa-speakers at Nambe, Pojoaque, Tesuque, Ohkay Owingeh (San Juan), San Ildefonso, and Santa Clara. All of these pueblos are located along the Rio Grande between Santa Fe and Taos, two of the state's major tourist destinations.

The people of New Mexico's northern pueblos have been the subject of scores of travel photographs, appearing on postcards, brochures, guides, and souvenir books, for more than a century. Throughout this time, they exercised some degree of agency over who and what was photographed in their own communities. What they had little control over was how their images were contextualised, interpreted, and used once they entered the discourse of the travel industry. This started to change in the 1980s, when the people of New Mexico's eight northern pueblos began to produce, individually and collectively, their own travel media. Besides issuing postcards and brochures, the tribes joined together in the creation of a major guidebook, the *Eight Northern Indian Pueblos Visitors' Guide*, which has been published annually since 1988.

As the Pueblo people of northern New Mexico produce and circulate photographic images that represent themselves and their communities to the travelling public, we ask to what extent has their imagery complicated and undermined conventional practices in Southwest travel photography. But first, we need to trace the earlier involvement of northern Pueblo people in New Mexican tourism and the history of its associated travel photography.

Travel Photography in a Colonising World

The tourist industry and its travel media unfolded in New Mexico during the last decade of the nineteenth century, a time that coincided with the intense colonisation of Pueblo peoples. It was a period when their country was populated by growing numbers of foreigners, when their land and water rights were under siege, when their children were removed to federal boarding schools, and when their traditional forms of governance were threatened. This was an era of intense and forced assimilation, where the United States government made every effort to destroy the Pueblos' economic, political, and cultural sovereignty (Dozier 1961, 1970; Spicer 1962; Sando 1992).

In the midst of this assault, another movement was afloat that extolled the independent character of the northern Pueblos and their *exotic* way of life. The region's growing and economically profitable tourist trade, built in large part on ethnic tourism, pushed to preserve and re-create aspects of Pueblo culture for the travelling public. Over time, the travel industry appropriated, in short colonised, many symbols of Pueblo life (for example, the Zia sun figure) for its own interests and purposes. It assumed control over much of the visual and written imagery that would define Pueblo people and their culture for the travelling public (Weigle 1989). Although not always on their own terms, people from many different pueblos became involved in the new and growing travel industry, most visibly as artisans, dancers at village ceremonies, or performers at tourist attractions. Less visibly, they were employed in domestic, railway, and construction work at sites that sustained the infrastructure of the region's tourism (Peters 1996). Even though this infrastructure has been in plain sight ever since, it is hidden in the world made visible to the tourist because it is symbolically unmarked and outside the role the travel media has cast Pueblo people.

Early Imaginings and Railway Travel, 1894-1934

Much has been written about the predominant place Pueblo peoples occupy in the visual and written texts that advertise and accompany travel in the American Southwest, and the role the Santa Fe, Atchison, Topeka Railway and its major concessionaire, Fred Harvey, played in shaping the discourse (Thomas 1978; McLuhan 1985; Weigle 1989; Babcock 1990, 1993; D'Emilio and Campbell 1991; Dilworth 1996; Weigle and Babcock 1996). Some of this writing applies to the northern Pueblos of New Mexico, but much of it does not because these pueblos were outside the commercial reach of Fred Harvey until 1926. During the early years of the region's travel industry, most of the northern Rio Grande pueblos were isolated, accessible only by horse-drawn coaches and wagons from railroad stops at Santa Fe and Taos, both of which were served by another major line, the Denver and Rio Grande. Many of the northern Pueblos who participated in tourism early on did so at sites removed from their villages, including Santa Fe in New Mexico and

Garden of the Gods or Manitou Hot Springs in southern Colorado. With the arrival of Fred Harvey's famous "Detours," which took travellers by motorised vehicles to sites in northern New Mexico, the villages of the northern Pueblos started to occupy a more prominent position in the region's tourism (Anonymous, c. 1930s; Thomas 1978). Simultaneously, artist colonies were developing in the towns of Taos and Santa Fe, drawing more visitors to northern Pueblo lands (D'Emilio and Campbell 1992; Dilworth 1996).

From the beginning, the visual marketing of tourism for the northern reaches of New Mexico and adjoining areas of southern Colorado was the work of travel businesses operating out of Denver, Colorado, or local souvenir outlets, such as the Candelario Curio Shop in Santa Fe, Taos Drug in Taos, or the Hidden Inn at Garden of the Gods in Colorado. Their combined output was much less orchestrated and sophisticated than the travel media associated with Fred Harvey, often taking on the appearance of photographic productions aimed at small town markets (Albers and James 1983, 1990; Albers 1998). Much of the work was issued on a small scale as real photos or printed on travel ephemera in black and white or sepia tones. Sometimes, the images appeared in tinted or hand-coloured formats, but these never matched the prolific output of views representing Fred Harvey's Southwestern world, which drew mostly on Pueblo villages near Albuquerque, Grants, and Gallup, New Mexico (Weigle and Babcock 1996).

The symbolic tone of the photography was also somewhat different. It employed many of the same romantic conventions found in Fred Harvey's travel discourse, which represented the Pueblo world as an ancient, pristine, and picturesque civilisation – America's equivalent to the "Biblical Orient" (McLuhan 1985; Dilworth 1996; Weigle and Babcock 1996). Yet, it also took on features associated with the epic discourse of America's Frontier West with its equestrian warbonneted warriors. As these two tropes were played out in early twentieth century travel photographs, they gave the visual imagery of New Mexico's northern Pueblos a distinct appearance.

Like their counterparts elsewhere in New Mexico and Arizona, the early travel media distinguished and distanced the northern Pueblos by wrapping their images in a language of exoticism and mysticism, but at the same time, making them appear familiar and approachable to tourists. The Pueblos were described, on the one hand, as a "curious" and "quaint" peoples with "mysterious ceremonies" and "weird customs," but on the other hand, as "civilised," "self-sufficient," "diligent farmers," "skilled craft workers," and "friendly to outsiders." As one early Curt Teich postcard (A34042*)* caption put it,

> The Pueblo Indian of bold, resolute, and determined character is picturesque anywhere and always. Mute, solemn and dignified generally, yet have the antics of children at the dance ceremonies. Indians who are neither naked nor poor were Weavers, Silversmiths, Farmers and Irrigators before the new world was discovered. Independent today and ask no aid from Washington.

In this era, travel promoters needed to assure tourists of the tranquillity and safety of the region to which they were travelling, especially when early tourism in the Southwest was only a few decades removed from the Apache's much maligned and feared struggles of resistance. In short, the Pueblos were a perfect model for a sight-seeing destination built around a quest for novelty and discovery in "other" worlds where the differences were novel but not entirely foreign (Albers 1995).

In the years before 1935, nearly every pueblo community in New Mexico offered the makings of a novel sight worthy of record in the travel media. Travel photography only needed to highlight and selectively embellish some of the more unique and picturesque sights. Certain images were quickly adopted by the media and widely conventionalised. In keeping with the enchanted mystique of the region, ceremonial dancing was the most popular subject picturing northern Pueblo peoples in early travel photography. The Matachines of San Juan, Corn dancers from Tesuque, and Turtle dancers at Taos were included among the scores of pictures depicting dancers and dances in the ceremonial life of the northern Pueblos. Because these pictures occupied a dominant place, males figured more prominently than females in early travel-oriented photography. Although Barbara Babcock (1990, 1993) claims that much of the travel discourse surrounding the Pueblos was a feminised one, this does not accurately apply to the northern Pueblos of New Mexico.

Early travel photographs also contained numerous views of activities from everyday life, whether these were staged for the camera or taken as candid photos. Besides various illustrations of food production, there were numerous pictures of craftswork from silversmithing to pottery making. Then, there were many less specific views depicting living quarters and village street life. In most cases, the images pictured in the travel media were ones that best represented northern Pueblo life as an idyllic, pastoral world.

The placement of northern Pueblo people in their home settings illustrated and shaped the prelapsarian imaginings of their life (see Dilworth 1996). Indeed, the vast majority of scenes picturing dances and domestic activities were taken in clearly marked village settings. There were also pictures from locations invented by tourist promoters. Manufactured sites, such as the Pueblo House at Manitou Springs in Colorado, were settings for travel photographs in which aspects of Pueblo life were exhibited and demonstrated as performance. The plaza in Santa Fe, especially the portal in front of the Palace of the Governors, was also a place where northern Pueblo people met tourists early on and a popular spot for taking photographs. Most of the views from popular tourist attractions were sold along side those taken in village locations, an association which no doubt reaffirmed the "authenticity" of the imitation (Weigle 1989; Street 1992).

One of the most distinctive characteristics about early travel photographs of the Pueblos in comparison to tribal nations from many other parts of North America was the association of the subjects with an identifiable surrounding or activity (Albers and James 1983, 1990; Albers 1995). Typically, subjects were depicted performing an activity, or else, they were identified with an action and its products. Women were depicted as bread bakers or potters with ollas, while men were pictured as silversmiths with the tools of their trade, dancers wearing ceremonial regalia, and leaders with the staffs of their authority. Whether in picture or caption, peoples' identities were not generally divorced from a specific source of labour, leisure or ritual, nor were they typically separated from a particular ethnic or geographic base. Usually, the subjects were not devoid of a context, even when it was contrived.

Early travel photos depicting the northern Pueblos were filled with images that followed many of the conventions associated with realist photography (Hunter 1987). The subjects appear to be absorbed in their activities and unaware of the camera's presence. In the Southwest, as in many other travel destinations worldwide, ethnographic realism was a preferred mode of representation in early twentieth century travel photography, in part, because tourism catered to an educated, urban and largely middle (or upper) class clientele who expected to see sights that were "real" and "genuine" (Edwards 1996). Travel pictures were made to be seen and read as credible representations of the native peoples they pictured, as transparent images of the subject's own lived-in world. That the pictures, including those staged at manufactured attractions were able to parade themselves as a slice of cultural reality speaks directly to the persuasive character of the rhetoric behind the realist representations of the time (Younger 1983; Dilworth 1996). Since most early tourists had only fleeting encounters with Pueblo people at public dances and craft demonstrations, their primary source of judgement came from the images the travel media fed them. There was little to contradict the "truth" of these images or the "authority" of those who produced them. Under the cover of a visualising rhetoric that had become conventionally accepted and legitimised in the traveller's own world, Southwestern media-makers were able to convince the public that it was possible to discover (or recover) the timelessness and otherness of Southwestern Indians in authentic ways. And, they were able to do so while disguising the origin of their productions and the colonising relationships these invoked (Albers 1995).

Realism was also evident in prevailing styles of portraiture. In most portraits, the subjects faced the camera directly. This style of posing was popular a century ago, not only in surveillance photography but for school, wedding and family pictures as well. It was a neutral composition, betwixt and between more ignobling and ennobling constructions (Albers and James 1988). When combined with generic captions, such as "Governor of Tesuque," "San Ildefonso Indians Painting Pottery," "Indian Dance at Santa Clara" or "Water Carriers, San Juan," its neutrality deflected attention away from the character of the people as individuals towards the categorical types they were purported to depict. In the context of its time, this style of portraiture buffeted other efforts to convince travellers not only to take the sights they were seeing seriously but to assure them they were "genuine" as well. Although more aesthetic and contrived forms of picture-making also existed, these were not the prevalent mode of representation during the formative years of travel photography in the American Southwest (Albers 1995).

1518 PUEBLO INDIANS AT TEWA HOUSE,
PHANTOM CLIFF CANON, MANITOU, COLO.

Figure 2.1   Pueblo Indians at Tewa House. Manitou, Colorado (c. 1920)

Beyond the core imagery, which promoted the vision of Pueblo peoples as a premodern peasantry, there was the image of the Southwest as a wild frontier of tribal renegades resisting American conquest (Dilworth 1996). Different from the civilised phantasm of Fred Harvey's Pueblo world, the frontier west was a heavily masculinised place with proud and courageous fighting men deserving of a heroic place in history (Albers 1998). Generally portrayed in epic or contrived styles of portraiture, most of the subjects were totally decontextualised and their pictures carefully orchestrated to recreate the familiar spectacle of a noble "race" of warriors and chiefs popularised in the Wild West Shows of the era (Albers and James 1984b). The Tewa of Santa Clara, who performed at two of Colorado's popular tourist attractions, Garden of the Gods and Manitou Springs, were frequently pictured on photographs wearing warbonnets and other Plains Indian style "regalia." So were the people of Tesuque who were shown performing their Comanche Dances, an old dance among the northern Pueblos which historically accompanied their trade with tribes from the southern Plains. In fact, the northern Pueblos adopted and developed their own style of feathered headdress long before tourists arrived on the scene. Although the generic image of the American Indian as a Plains Indian did not escape early image-making in the Southwest, it never overpowered it in the way it did in many other regions of North America (Albers and James 1984a).

Modern Movements and Automobile Travel, 1935-1975

After 1935, when the heart of New Mexico tourism moved to regions north of Albuquerque, the northern pueblos became the most important indigenous players in the state's travel industry – a shift that is clearly revealed by their increasing and eventually dominating presence in travel pictures. The frequency of their images on postcards, as one example, now dramatically outpaced New Mexico's Zuni, Isleta, and Keres peoples whose images had figured so prominently in the early tourist-oriented pictures of Fred Harvey (Albers 1995).

As the twentieth century progressed, much of what had been distinct about Pueblo cultures was no longer subject to the tourist's gaze. This happened as Pueblo communities selectively "compartmentalised," using an expression from Edward Dozier (1961), the ordinary and unique aspects of their lives that were closed from those open to outsiders. In time, many Pueblo communities and ceremonies were off-limits to the public and the camera alike. It also occurred as a result of the changing appearance of the communities themselves, many of which began to look indistinguishable, at least externally, from much of the rest of the Southwest. As Pueblo people took on more of the fashions and material conveniences of the wider population, they were less likely to present the kind of unique and exotic "look" extolled in local tourism.

In order to keep the enchanted and enduring Southwest alive, travel photography was enlisted to help preserve the myths. Even more than before, photographers focused and framed their pictures to conform to the lore and legend so successfully popularised and conventionalised in the earlier travel productions of Fred Harvey. Over time, and through a gradual process of selection, the mythical "look" of the Pueblo Southwest was sustained by a narrow range of pictorial subjects and styles, endlessly replicated in the work of different photographers. One apt example is the image of women baking bread in the beehive-shaped ovens the northern Pueblos call a *horno* or *panteh*. Whether pictures of this activity come from Isleta and Cochiti in 1910 or Taos and Tesuque in 1940, they follow nearly identical stylistic conventions. All of them depict their subjects wearing a traditional *manta* and shawl. Staged or taken from "real life," this sort of stock image represents a classical embodiment of the prelapsarian imagining of the Southwest.

Photography is especially well suited to shaping images into myths because its products are long-lived. Barring mutilation, destruction, or other alteration, photographic negatives and prints retain the integrity of their constituted appearances well beyond the time of their original making (Sekula 1990). This self-evident aspect of their material existence had important consequences for perpetuating ahistorical notions about Pueblo life. From 1920 until 1950, for example, the travel media in the Southwest recycled many photographs taken at the turn of the twentieth century. One photograph of a potter from Tesuque, for example, was reproduced continuously on the picture postcard for nearly half a century. In its later manifestations, it was sandwiched between more modern views without any reference to its historic origins, perpetuating the idea that Pueblo peoples, like their photographic appearances, were frozen in time.

With the arrival of commercial kodachrome processing after World War II, a totally new line of photos was created for the tourist market. The new technology offered travel photographers an opportunity to remake the Southwest's visual myth with a more updated look. Yet, most of the newer photos did not depart significantly in their subject matter from the older ones. The pastoral myth persisted, and Pueblo people remained its quintessential representatives. Photography now focused its sights on the most exotic production sites, such as outdoor bread baking, but most of all on the manufacture of pottery and other crafts sold in tourist markets (Albers 1995). The multitudinous images of potters and their wares, not only enhanced a desire and demand for objects of "exotic" manufacture, but they helped to authenticate their status as pre-industrial artefacts as well. By placing these pictures amidst scores of others, which depicted the unique ceremonies, settlements and domestic activities of Pueblo people, the extraordinary status of these objects and their function as travel fetishes became secure (Parezo 1990; Dilworth 1996).

As in the past, photography continued to support the mystique that surrounded Pueblo ritual and ceremonialism. Many of the Pueblo communities, commonly depicted in travel photographs in earlier times, now prohibited picture taking at their ceremonies. The northern Pueblos, however, continued to permit amateur and commercial photography at select village events. As a result, the vast majority of modern ceremonial dance views taken at village locations came from the Tiwa at Taos and the Tewa at Tesuque, San Ildefonso, San Juan, and Santa Clara. An even larger number of contemporary dance images came from popular tourist attractions such as Gallup, New Mexico's Intertribal Ceremonials, which, after the mid-1930s, became the single most important source of travel photographs, picturing dances and dancers in the Southwest.

Even though the dancing performed for tourists at Gallup was decontextualised, part of a pay-for-view, public spectacle, its exotic and enchanted origins still needed to be sustained. One way this was done was to give the images and their accompanying texts a dramatic flair. Unlike the past, when pictures showed dances in progress, the ones taken after World War II depicted the dancers in statuesque poses or in staged imitations of various dance performances. They were also captioned with superlatives of the "exciting," "thrilling," "spectacular," and "magnificent" (Gallup Intertribal Ceremonial Programs 1932-1961). In the end, the image/text of these pictures kept the myth of the Southwest alive, but it did so in ways that appealed to the sense of excitement which surrounded the extravaganzas of the day.

Realist styles of photography continued, but there was a growing trend, especially after World War II, to pose Pueblo people in more ennobling ways. Epic styles of portraiture became much more common (Albers and James 1988). Here, subjects were posed in statuesque stances, dressed in regalia for festive and ceremonial occasions. They were positioned in the foreground of a picture at or above the level of the camera's eye with backgrounds dominated by sky or marked symbolically by picturesque attractions and landscapes. In many ways, this more affected style of posing over determined the extraordinary character of the subjects, especially in settings, such as Gallup, where there was little to distinguish the world of the tourists and the native performers because it was the same.

Paralleling this transition was a change in the tone of the written discourse. Although not without their own stereotypes and superlatives, many of the older texts were intended to be literal or descriptive in a referential sense rather than figurative. From the 1940s to the 1970s, travel publications were filled with hyperbole and largely dominated by epic-making texts as in this caption, describing Taos in the late 1940s,

> A land utterly alien and astonishing ... The old traditions and religious customs, so arrogant, so proud, so magnificently are they clung to, that the pueblo dwellers merit a place among the great peoples of the world (Southwest Post Card Company #37).

The representation of the Pueblos through an epic discourse was nothing new, of course, but the visual and narrative devices that were used to express it became much more dramatic and affected. Generally speaking, even though changes were taking place in the ways in which Pueblo traditions were being represented by the travel media, the essential character of the trope that defined them not only persisted but it became even more firmly established. Under the enduring spell of an enchanted and exotic Southwest, other imaginings were diminished or re-invented in the image of the dominant myth.

One of these was the trope of the "frontier" West, which was typified in the image of the chief wearing a warbonnet. Northern Pueblo men were frequently attired in this regalia when they posed for pictures at popular tourist attractions and even in their home settings. In one sense, they became the Southwest's "stand-in" for the generic image of the American Indian. The growing popularity of the Hollywood movie and TV Western during this era also bolstered the strength of the frontier legend in the public mind, and as a result, the imagery associated with it had growing appeal in Southwestern tourism as it did in many other parts of the United States (Albers and James 1983; 1985a).

Equally important was the growth of the modern Plains powwow among tribes in the Southwest. Beginning in the 1950s, many Pueblo families started to participate in powwows as a social activity. Some men, many of whom came from Taos Pueblo, became prize-winning fancy dancers and popular subjects in local travel photographs. At least from the vantage point of the travel industry, the pictures of these dancers with their roaches and elaborately feathered bustles satisfied the needs of the popular frontier trope with its warriors and chiefs. Even though the imagery supporting this trope became more widespread in travel photography, especially during the 1950s and 1960s, it still never replaced the Southwest's own homegrown legends. Instead, it was melded to the region's dominant myth structure and ultimately "civilised" by its association with the

Pueblos whose peaceful, pastoral reputation overwhelmed the "wild" frontier imaginary generally linked to the cultures of Plains Indians.

By the 1960s, Southwestern tourism was organised almost entirely around automobile travel and the highly diversified attractions and agencies that supported it. The businesses that offered accommodation and other infrastructural support for tourism were now generally separate from those involved in the production of travel ephemera, although state-run agencies, such as the New Mexico Department of Development, provided a structural link between the two. In New Mexico, many of the pictures sold commercially on postcards, souvenir folders, and placemats represented the work of photographers who were associated with the state's official tourist agency and whose pictures appeared regularly in the state's travel-oriented publication, *New Mexico Magazine*.

Sight-seeing travel in the American Southwest was also now affordable to a wider segment of the American public, and as parodied in Chevy Chase's film, "Vacation," it became a family affair, a kitsch bound highway excursion with Disneyland (and its adult counterpart, Las Vegas) the ultimate destination for mythic renewal. Neon lit kachinas, faux tipi motels, larger than life statues of warbonneted chiefs, and tacky plastic, tom-tom souvenirs dotted the Southwest's travel landscape. Travel brochures still extolled the distinctiveness, timelessness, and beauty of the region, but tourists were beckoned to the area through a different sort of code. Enlightenment was no longer the principal lure. Now, the enticement was entertainment – a search for excitement, the thrill of the spectacle, which events like Gallup's Intertribal Ceremonials willingly obliged. Even the intrinsic features of the Southwest's cultural landscapes were no longer sufficient to attract visitors. Instead, tourists were lured by sights described as even more extraordinary and by sites unabashedly set in a world of fantasy and fun (Weigle 1989).

Much of what we've described, thus far, is not unique to tourist photography in the American Southwest, but instead it reflects broader trends in the popularity and use of certain photographic conventions in travel photography worldwide. Like the more general tourist productions of which they were an essential part, travel photographs came to draw increasingly on epic forms of image-making, which not only homogenised ethnic subjects according to singular and highly stereotyped models, but which also increasingly situated them in a ritualised symbolic discourse outside the course of their ordinary life (MacCannell 1976; Albers and James 1983, 1987, 1988; Urry 1990; Dann 1996; Edwards 1996; Nash 1996: 59-78).

Post-Modernism and Global Travel, 1976-Present

By the early 1980s, another shift was well underway in Southwestern tourism and the photographic productions that supported it. Select scenic destinations, such as Taos Pueblo, now a World Heritage Site, became meccas for a growing international tourist trade whose interests were different from the "tourist-trap" vacationers of the 1950s and 1960s. Hosting travellers from all points of the globe, the Southwestern travel industry started to refocus its sights once again towards the high end of the tourist market. Over the next twenty years, a growing segment of the industry began to cater to the "discerning" tastes of a wealthier, international clientele, although the earlier kitsch marketplace remained very much alive as substance and parody along Route 66. The new "designer" tourism gave rise to a group of small, independent photographers and publishers whose work typically avoided the tourist ready attractions of an earlier time. The search for enlightenment and entertainment still marked the travel industry's call but now much of it was built around a nostalgic search for a distant colonial past rather than any actual encounter with the people who stereotypically and symbolically embodied ethnic "otherness" in the Southwest.

The modernist travel sensibility with its exaggerated "exotic" tropes still continued, but now it was produced in a world where most Pueblo people were no longer willing to submit themselves and their images to the travel industry's conventionalised turf. Today, New Mexico's northern Pueblos have strict rules governing the behaviour of outsiders visiting their villages – including the ones carrying cameras. Each pueblo sets its own regulations for commercial and amateur photography. Depending on the time of year, type of dance, and the elected officials in charge of the pueblo, taking photographs at the villages and/or at their dances can be open or off limits to the public. Visitors are required to inquire about local policies at tribal government offices before they travel to the villages. Importantly, outsiders are no longer able to photograph at will.

Commercial travel photography in the Southwest has adapted to these restrictions in several different ways. One way, as in the past, is to recycle its images. Kodachrome pictures, taken fifty years ago, still appear, but now they are repackaged with new frames to give the illusion of being present in the time and experience of the tourists who consume them. In Figure 2.2, the perennially popular bread bakers return.

When modern images of Pueblo people no longer offer up exotic enough sights for the grist of the travel industry's visual mills, vintage pictures feed the mythic icons. In the late 1980s, the appearance of early twentieth century photographs on postcards, calendars, and other travel ephemera became part of a growing trend in the region's visual productions. Especially popular were photographs taken by Edward S. Curtis that played into the mythical tropes of the Southwest as an enchanted and enduring place. And in some respects, these pictures performed the service of symbolic icons even more decidedly than many contemporary photographs because of their highly stylised and romanticised compositions (Lyman 1982).

Also, many travel photographers in the Southwest began to represent the region's ethnicity more through its physical sites and objects than its people. Today, Pueblo people are less likely to be present in the photographs, which picture their villages. Taos, as one example, remains one of the most popular sites in New Mexico for travel photographs, but while the architectural features of this

BAKING BREAD

INDIAN WOMAN

Figure 2.2   Taos Pueblo woman

pueblo are widely pictured, the number of those, which include the residents of this village, has declined dramatically.

Some villages became off-limits entirely to outside photographers. Santo Domingo, a Keres village, south of Santa Fe is well known for its conservatism and its strict policies prohibiting photography. This village, however, is the site of one of the most spectacular summer Corn Dances open to the public and a popular destination in the region's ethnic tourism. Undaunted by the prohibitions, a few local photographers took aerial views of the pueblo during a summer Corn Dance in the 1960s. Much to the dismay and consternation of the village, some of these photographs were published on postcards. Today, an entire sequence of aerial photographs of the pueblos in New Mexico has been issued, including the communities who still prohibit commercial photography in their villages. Locations that are off-limits on the ground, such as the cemetery at San Juan, are in plain sight. Notwithstanding their own policies, the Pueblos of New Mexico are still subject to unwanted surveillance from the eye of the region's travel media.

In the 1980s, travel photographers also focused their sights on aesthetic displays of rugs, baskets, pottery, and jewellery instead of the artists behind their production. Even more than in the past, modern tourism requires the source of its invention to be obfuscated. "Genuine" Pueblo art, which is now often produced with the assistance of mechanised tools, can no longer be visualised in extraordinary ways except as a form of re-enactment, and this is different from the past when photography was still able to capture exotic images of artisans at work. Even though the artisans are absent in much of the region's contemporary travel photography, their arts and crafts are still singled out at the travel industry's altar of curios and relics and illustrated as such (Albers 1996: 248-250).

It is not difficult to argue for the existence of a hegemonic and colonised travel discourse in which Pueblo peoples and their cultures became hermetically sealed in photographic images that made them appear, according to one 1960s New Mexico travel brochure, exactly the same as they did when Coronado came to the area in the 1540s (Anonymous, c. 1950s). There is no doubt that the travel industry in New Mexico appropriated Pueblo culture for its own purposes and interests. Nor is there any question that in its various productions it objectified Pueblo people by mystifying their life and transforming it into an historic relic (Weigle 1989; Babcock 1990, 1991; Dilworth 1996; Weigle and Babcock 1996). Photography plays an important role here, in so far as the appearances of its products carry ambiguous meanings. It is easy to voice over a photograph and give it meaning that has nothing to do with the subjects it represents (Berger 1972, 1983; Sontag 1973; Hall 1982; Hunter 1987). In the Southwest, as in other travel destinations worldwide, the imagined "truth" of an image is authenticated not so much in reference to the subjects it pictures, but rather in relation to the context of its circulation as a commodity in the region's tourist trade (Cohen 1988).

Like other manufactured simulacra (Baudrillard 1993), photographic images saturate the travel landscape of New Mexico. Severed from the original contexts of their making and the lives of the people they depict, popular travel pictures are easily reinvented across a variety of different iconographic terrains. The same pictures are reprinted on millions of copies in travel guides and souvenir books and on posters, brochures, postcards, and tablemats. They are sold everywhere from motel lobbies and highway cafes to museums and trading posts. The very redundancy of certain pictures on postcards and other tourist stuff, coupled with the magnitude of their commodified production, dwarfs other windows for seeing and interpreting the people they picture. It is little wonder, then, that such pictures appear hegemonic and became the standard against which the visual "reality" and "authenticity" of Pueblo people gets judged by the travelling public.

Lacking alternative imaginings, much of the Southwest's modern travel industry has become trapped by its own mythical productions, unable to escape their ultimate consequences except as a form of parody. In fact, some of the absurd outcomes of its gaze have been lampooned on comic cards released by the industry over the past fifty years. The problem the industry now faces is that it no longer occupies a position where it can seamlessly control and manipulate the images behind its ethnic productions as it did in earlier decades. The subjects of the imagery now have a voice and the means to complicate and challenge the myths the travel industry uses to represent them.

Figure 2.3   San Juan Pueblo women baking bread
Source: Courtesy of Eight Northern Pueblos Council, Inc.

Towards a De-Colonising Travel Photography

Over the past twenty years, the northern Pueblo people of New Mexico have
started to establish their own travel and recreational venues, many of which depend
only tangentially on selling their "ethnicity" either as a product or a performance.
A golf resort at Pojoaque, a ski resort at Taos, and casinos at Ohkay Owingeh,
Santa Clara, and Tesuque offer Pueblo people an entrance into the region's travel
industry not defined by their historic and traditional relations to tourism. Even
where their lives are still affected by ethnic tourism, as in art and crafts production
or in the popular public dances at their villages or at outside exhibitions, they
now have considerably more power to define the terms of their relationship to the
travelling public.

It is not coincidental that New Mexico's northern pueblos launched their own travel guide in 1988, the same year the Indian Gaming Regulatory Act was passed. More than anything else, the passage of this legislation offered the Pueblos an opportunity to participate in the lucrative travel market on their own terms and with no outside competition. From its inception, the *Guide* was intended not only to advertise the tribally owned casinos and resorts, but also to educate the public about Pueblo culture and history.

The publication of the *Visitors' Guide* is completely controlled and supervised by a council whose members are appointed by the governors of each of the eight northern pueblos. Most of these members either live in the pueblos and/or work for their tribal governments. Along with staff hired or contracted to produce certain articles and photos, the council organises the general layout of the text and its accompanying visual imagery, which is then submitted to the governors for their final approval.

The *Guide* is widely circulated at tourist accommodations and destinations throughout New Mexico, and it serves as a major alternative source of information on the northern pueblos and their people. The *Guide* lays out the proper conduct for outsiders visiting the villages and taking in the public ceremonies, including the use of the camera. It also contains many different essays written by members of the eight northern Pueblos. These include descriptions of each pueblo's landscape, its history, political traditions, ceremonial life, artistic contributions, and economic aspirations. Every *Guide* is liberally illustrated with photographs, and most of the contemporary photos represent the work of photographers who also come from one or more of the eight northern pueblos.

Hundreds of different photographs have appeared in the *Guides* over the past fifteen years. Many of them follow the same visual conventions in other contemporary travel publications. At first glance, it would appear that Pueblo people are perpetuating the popular mythic tropes that have represented them for so long in the mainstream travel industry. On closer inspection, it becomes clear that the picture is more complicated. This is true in several different ways.

Like other contemporary travel publications (Gattuso 1991; Gibson 2001; Cheek and Fuss 1993; Tremble 1993; King and Greene 2002) about New Mexico and the greater Southwest, the most popular and recurring pictures in the northern Pueblo *Guide* are images of dances and dancers. The Buffalo and Comanche Dances are the two most frequently pictured in the guides, followed by Deer, Cloud and Corn Dances. The first two dances are invariably public dances in New Mexico's northern pueblos and open to all visitors. These two dances are also the ones most frequently performed, along with Eagle and Butterfly Dances, at attractions such as Gallup's Intertribal Ceremonial and the Eight Northern Indian Artists and Craftsman Show. These dances, Buffalo and Comanche, almost always allow photography. The other dances pictured in the *Guide* are not performed as commonly outside the villages, and there are differences among the pueblos in the appropriateness of opening them to public spectatorship either within or outside their own village setting. As would be expected, the policies that govern taking pictures of these dances also vary.

In contrast to some other travel productions (Gibson 2001; Cheek and Fuss 1996), the producers of the *Visitors' Guide* are very discriminating about what kinds of dance imagery they include in their publication. Pictures of certain dances and dancers, whether based on historic or modern photographs, are not included. The *koshare* dancers are considered sacred, and their contrary behaviour, which is often interpreted by outsiders as "clowning," is serious business not a form of playfulness. Almost always, they appear in dances that are off limits to photography. Since the *koshare* are no longer subjects of the contemporary camera's eye, the travel media continue to use historical photographs to picture these popular figures – much to the chagrin of Pueblo people.

As in other guidebooks and travel publications, the emphasis on ceremonialism has always distinguished the Pueblos and given them a distinct cachet in the panoply of images that mark their presence in the region's travel photography. With few exceptions (Trimble 1993), most of the publications (Gibson 2001; Cheek and Fuss 1996; King and Green 2002) produced by the mainstream media use the dances and dancers to illustrate and reinforce the old myth of the Pueblos as an unchanging and exotic peoples, still clinging to ancient practices from another time and place. In the northern Pueblos' *Guide*, by contrast, the dances are described as traditions which, although old, are not relics of some soon to vanish past, but an integral part of the way in which Pueblo peoples renew their place in the present.

Many of the other stock subjects in the region's travel photography, notably bread-baking and pottery-making also appear in the *Guide*. In the northern Pueblo *Guide*, the first activity has been visually represented in two ways. One cover of the *Guide* for 1995, for example, illustrates bread-baking in a manner not dissimilar to its appearance in other travel productions, but another, in 1994, depicts it in a way more faithful to how people actually appear when the activity is undertaken.

In most northern Pueblo communities, bread is baked in the old fashioned way only for ceremonial occasions, although Taos women often produce and sell it for the tourists who come to their main village. Unlike its conventional representation in travel photographs, this is not a solitary activity. It involves groups of women working together, often with the assistance of men and children. It is also an activity people usually carry out in their everyday dress, not the festive regalia typically pictured in travel pictures. The kind of imagery that is more "true to life," so to speak, rarely appears in the mainstream travel media, but it is commonplace in the publications of the northern Pueblo.

Importantly, while the northern Pueblo *Guide* includes conventional travel photos, it does so in conjunction with other images that picture people in settings and contexts more typical of their actual lived-in experience. In fact, one *Guide*, the one for 2002, contained family snapshot pictures from each of the pueblos. Photos of family groups have a long history in the mainstream travel press, part of the familiarising strategies that make subjects appear approachable and receptive

Figure 2.4   Pueblo of Tesuque corn dance

Source: Courtesy of Eight Northern Pueblos Council, Inc.

to a tourist's gaze (Edwards 1996). The message of familiarity in conventional travel photographs, however, is surrounded by appearances that communicate a distance between the subject and the viewer. This is achieved either through the exotic dress and settings in which the subject appears, the formality and polished character of the poses they assume, and/or the generic nature of the titles that caption them. The *Guide's* candid and named family snapshots engage a sense of intimacy all around, but they do so for a different reason. By presenting their own people through a familiar and candid photographic format, the creators of the *Guide* want their viewers to see Pueblos as real people not cultural icons who stand passive and voiceless in the face of the traveller's gaze. In Pueblo travel productions, the images of families and their everyday lives are used to humanise the subjects not to further objectify and distance them.

One can argue that the preponderance of photos of children in the *Guide* has similar implications. In fact, the most noticeable departure of Pueblo travel publications from the mainstream media is the extent to which children dominate the imagery (see Figure 2.4).

Pictures of children also appear in mainstream travel productions, more so than in the past, but they are generally depicted wearing their feast day or dance

Figure 2.5   Pueblo of Picuris schoolchildren

Source: Courtesy of Eight Northern Pueblos Council, Inc.

regalia. In the northern Pueblo *Guide*, children are pictured this way too, but they are also seen riding bicycles around their villages, boarding school buses, and playing sports (Figure 2.5).

The written texts that surround these images convey, once again, that Pueblo people are not relics of some dying or vanishing past; they are very much a part of the present and the future. It is for the future well-being of their children, one of the *Guide's* advertisements from 1998 states, that economic development is being pushed in the travel and recreational industry. Elsewhere, children are described as the ones who will carry Pueblo traditions into the future and keep them alive for generations to come.

That Pueblo people are as much about the present and the future as they are about the past is a theme that recurs throughout the *Pueblo Guide* and underlies much of its visual and written discourse. Notwithstanding the recurring message that the Pueblos are living peoples not ancient cultural icons, those who produce the *Guide* are not above using historic pictures or taking modern photos in imitation of a popular vintage images. One photograph (Figure 2.6), taken at the waterfalls near Nambe Pueblo, which appeared in the 1999 *Guide*, is clearly an imitation of a work by Edward S. Curtis.

Figure 2.6   Offering at Nambe Falls

Source: Courtesy of Eight Northern Pueblos Council, Inc.

Here, the Tiwa and Tewa people of New Mexico are laying claim to one of the tropes that stereotypically represents them in the region's travel media. This appropriation should come as no surprise. After all, Edward S. Curtis' style of photography has great appeal both inside and outside Indian country. Its ennobling subjects and stylised conventions reinforce aspects of the past from which many contemporary Pueblo people draw strength and inspiration. They invoke a nostalgia and sentimentality that portrays their ancestors in a respectful and aesthetically pleasing fashion. In the Pueblo context, these images are not icons, invoking a strange yet pleasurable otherness – the role they prominently play for the population at large (Lyman 1982). Instead, their iconic meanings engage a sense of continuity, an unbroken connection with their cultural past, which, although idealised, still has the possibility of manifesting itself in the present.

The Nambe picture is also important because it speaks to an important value that is very much alive in today's Pueblo world, and that is, respect for the natural world. This and other similar images express the continuing importance of certain traditional values. Importantly, photos such as this one are not created to appeal to the New Age variety of tourism currently popular in the Southwest, with its mysticism, sacred site tours, and vision quest packages, even though they may very well do so. When the postcard version of the older Curtis picture is sold on postcard racks today, the meaning of the image to Pueblo people is lost. It stands alone, without any written text to position it, and as a result, it is easily read and interpreted through the other commodified tropes that surround it, including the New Age varieties. When its mimesis appears in the *Guide*, it is not severed from a lived-in time and place. It becomes very clear from the text and other pictures that this is a symbol from the Pueblos' past that lives on in their present. Even though the image is still situated within an arena of commodity production, the setting is controlled by native voices and visions.

More than anything else, the northern *Pueblo Guide* reveals that Pueblo people are now taking charge of the some of the images which have defined them in the travel media and assigning them new meanings. For the most part, the producers of the *Guide* have not separated themselves, at least visually, from tropes promulgated by the travel industry at large, nor have they reimagined their communities through strikingly different visual subjects and compositions. One of the reasons, perhaps, for the persistence of the old conventions in the Pueblo controlled media is that many of the subjects they feature are ones Pueblo people take pride in. While the village dances and ceremonies have been a subject of tourist spectatorship for more than a century, they still belong to the Pueblo people. The interpretation of the dances pictured in travel photography may have become distorted, but the photos themselves are "real." They depict actual events that Pueblo people enact for their own purposes, and as such, the images of them can be recovered and assigned meanings that are authentic to the time, place, and people pictured. An important part of the Pueblo effort is to put their own unique stamp on popular travel productions, and in the process, to demystify the conventional images and offer travellers alternative ways to interpret them.

Over the years, northern Pueblo people have negotiated with outsiders, through countless individual and community decisions, what they would permit them to see and capture on camera. By doing this, they helped shape the appearance of the imagery the public sees and does not see. In keeping with other processes of compartmentalisation, which have long characterised their encounters with outsiders, Pueblo communities have been able, on the one hand, to hide and protect certain aspects of their own cultural world from outside pressure and influence. Yet, on the other hand, they have been able to share and exchange those aspects of their culture whose integrity is not threatened by its positioning in a world of spectatorship and commerce. The Pueblos have always had agency and ownership over their dances as carried out in their own villages, but now they are extending their proprietary interests even further. By redefining and reimagining what the travel images of them mean, they are asserting control not only over what the tourist sees but also how they come to interpret what they see.

This is a step towards a more revolutionary transformation where Pueblo people begin to change the content and composition of the images through which the travelling public sees them. This is already taking place in the *Guide*, but it is especially apparent in the work of many contemporary Pueblo photographers

(Masayesva, Jr. 1983; Younger and Masayesva, Jr. 1983: 41-109; Sakiestewa 1992; Hubbard 1994) whose work typically appears in fine art venues rather than tourist markets. The few images of this sort that appear on travel ephemera picture conventionalised subjects but they do so with an ironic twist. A postcard published by the Keres Pueblo of Zia, near Albuquerque, pictures traditional dancers, drinking Coca-Cola and resting against a vintage car. It draws the viewer to the subjects through the conventionalised appearances it pictures but it also destabilises the gaze with the juxtaposition of non-conventional objects. In this kind of picture, modern Pueblo photographers force tourists to see their people in ways that simultaneously break down and reaffirm the separateness of their world from the tourist's. The reaffirmation is not about authenticating some exotic "otherness," but about honouring and validating the special and unique place of Pueblo peoples in the modern world. Pueblo travel photographs are not intended to recreate images that objectify their subjects and turn them into relics without history. Their photography is about dynamic human spaces, where people live, work, and celebrate and where the present is the past and the future. Even though this photography may draw on some of the stock conventions found in the region's travel photography, the intent is different – it is life-affirming.

References

Albers, P.C. (1995), "Enchanted, Enduring and Exotic Images of Southwestern Indians in Photography," Unpublished Paper Presented at Imaging The West Symposium, Autry Heritage Museum, Los Angeles.

Albers, P.C. (1996), "From Lore to Land to Labor: Changing Perspectives on Native American Work", in Littlefield, A. and Knack, M. (eds), *Native American Wage Labor*, University of Oklahoma Press, Norman, pp. 245-273.

Albers, P.C. (1998), "Symbols, Souvenirs and Sentiments: Early Postcard Imagery of Plains Indians", in Geary, C. and Webb, V. (eds), *Delivering Views,* Smithsonian Press, Washington, D.C., pp. 131-167.

Albers, P.C. and James, W.R. (1983), "Tourism and the Changing Image of the Great Lakes Indian", *Annals of Tourism Research*, Vol. 10, pp. 128-148.

Albers, P.C. and James, W.R. (1984a), "The Dominance of Plains Indian Imagery on the Picture Postcard", in Horse Capture, G. and Ball, G. (eds), *Fifth Annual 1981 Plains Indian Seminar in Honor of Dr. John Ewers*, Buffalo Bill Historical Center, Cody, pp. 73-97.

Albers, P.C. and James, W.R. (1984b), "Utah's Indians and Popular Photography in the American West: A View From the Picture Post Card", *Utah Historical Quarterly*, Vol. 52, pp. 72-91.

Albers, P.C. and James, W.R. (1987), "Tourism and the Changing Image of Mexico", unpublished Paper presented at the meetings for the Society for Applied Anthropology, Oaxaca, Mexico.

Albers, P.C. and James, W.R. (1988), "Travel Photography: A Methodological Approach", *Annals of Tourism Research*, Vol. 15, pp. 134-158.

Albers, P.C. and James, W.R. (1990), "Private and Public Images: A Study of Photographic Contrasts in Postcard Pictures of Great Basin Indians, 1898-1919," *Visual Anthropology*, Vol. 3, pp. 343-366.

Anonymous, New Mexico (n.d.), *Harveycar Motor Cruises: Off the Beaten Path in the Great Southwest*, Undated travel brochure, c. 1930s. Collections of the Museum of New Mexico.

Anonymous, New Mexico (n.d.), *New Mexico 'Land of Enchantment.'* Undated travel brochure, c. 1950s. Collections of the Museum of New Mexico.

Babcock, B. (1990), "'A New Mexican Rebecca': Imaging Pueblo Women," *Journal of the Southwest*, Vol. 32, pp. 400-437.

Babcock, B. (1993), "Bearers of Values, Vessels of Desire: The Reproduction of the Reproduction Pueblo Culture," *Museum Anthropology*, Vol. 17, pp. 43-57.

Baudrillard, J. (1993), *Symbolic Exchange and Death*, Sage Publications, London.

Bennett, R. W (1934), "Which Road to the Pueblo?," *New Mexico Magazine*, Vol. 11, pp. 11-13.

Berger, J. (1972), *Ways of Seeing*, Penguin Books, London.

Berger, J. (1980), *About Looking*, Pantheon Press, New York.

Cheek, L. W. and Fuss, E. (1996), *Santa Fe*, Compass American Guides, Fodor's Travel Publications, Inc, New York.

Cohen, E. (1988), "Authenticity and Commodization in Tourism," *Annals of Tourism Research*, Vol. 15, pp. 371-386.

Dann, G. (1996), "The People of Tourist Brochures," in T. Selwyn (ed.), *Myths and Myth Making in Tourism*, John Wiley & Son, New York, pp. 33-47.

Dilworth, L. (1996), *Imagining Indians in the Southwest: Persistent Images of a Primitive Past*, Smithsonian Institution Press, Washington, D.C.

Debord, G. (1980), *Society of the Spectacle*, Black and Red Press, Detroit.

Dozier, E. (1961), "Rio Grande Pueblos," in E. Spicer (ed.), *Perspectives in American Indian Culture Change*, University of Chicago Press, Chicago, pp. 94-186.

Dozier, E. (1970), *The Pueblo Indians of North America*, Holt, Rinehart and Winston, New York.

Edwards, E. (1996), "Postcards: Greetings from Another World," in T. Selwyn (ed.) *Myths and Myth Making in Tourism*, John Wiley & Son, New York, pp. 197-221.

Eight Northern Indian Pueblos Council, Inc. (1988-2003), *Eight Northern Indian Pueblo Visitors' Guide*, Ohkay Owingeh (formerly San Juan Pueblo), New Mexico.

D'Emilio, S. and Campbell, S. (1990), *Visions and Visionaries: The Arts and Arts of The Santa Fe Railway*, Peregrine Smith Books, Salt Lake City.

Gallup Intertribal Ceremonial Association (1924-1961), *Programs*, Collections of the Museum of New Mexico.

Gattuso, J. (1980), *Insight Guides: Native America*, Hoffer Press Ptc. Ltd., Singapore.

Gibson, D. (2001), *Pueblos of the Rio Grande: A Visitor's Guide*, Rio Nuevo Publishers, Tucson.

Graburn, N.H.H. (1961), "Tourism: the Sacred Journey," in V. Smith (ed.), *Hosts and Guests*, University of Pennsylvania Press, Philadelphia, pp. 17-31.

Hall, S. (1982), "The Rediscovery of "Ideology": Return of the Repressed in Media Studies," in M. Gurevitch, M., Bennett, T., Curran, J. and Wollacott, J. (eds), *Culture, Society and The Media*, Methuen, London, pp. 56-90.

Hollingshead, K. (1992), "'White' Gaze, 'Red' People – Shadow Visions: The Disidentification of 'Indians' in Cultural Tourism", *Leisure Studies*, Vol. 11, pp. 43-64.

Hubbard, J. (ed.) (1994), *Shooting Back from the Reservation*, New Press, New York.

Hunter, J. (1987), *Image and Word: The Interaction of Twentieth-Century Photographs and Texts*, Harvard University Press, Cambridge.

King, L.S. and Greene, G. (2002), *New Mexico for Dummies: A Travel Guide for the Rest of Us!*, Hungry Minds Inc., New York.

Laxson, J. (1991), "How 'We' See 'Them:' Tourism and Native Americans", *Annals of Tourism Research*, Vol. 18, pp. 365-391.

Lyman, C. M. (1982), *The Vanishing Race and Other Illusions: Photographs of Edward S. Curtis*, Pantheon Books, New York.

MacCannell, D. (1976), *The Tourist: A New Theory of the Leisure Class*, Schocken Books, New York.

Masayesva, V. Jr. (1983), "Kwikwilyaqa: Hopi Photography," in E. Younger and V. Masayesva, Jr. (eds), *Hopi Photographers/Hopi Images*, Sun Tracks/ University of Arizona Press, Tucson, pp. 9-12.

McLuhan, T.C. (1985), *Dream Tracks: The Railroad and the American Indian 1890-1930*, Harry N. Abrams, New York.

Nash, D. (1987), *Anthropology of Tourism*, Pergamon Press, New York.

Parezo, N. (1990), "A Multitude of Markets," *Journal of the Southwest*, Vol. 32, pp. 563-575.

Peters, K. (1996), "Watering the Flower: Laguna Pueblo and the Santa Fe Railroad, 1880-1943," in A. Littlefield and M. Knack (eds), *Native Americans and Wage Labor: Ethnohistorical Perspectives*, University of Oklahoma Press, Norman, pp. 177-198.

Pomeroy, E. (1957), *In Search of the Golden West: The Tourist in Western America*, Alfred A. Knopf, New York.

Sakiestewa, R. (1991), "Weaving for Dance," in L. Lippard (ed.), *Partial Recall*, New Press, New York, pp. 73-76.

Sando, J.S. (1992), *Pueblo Nations: Eight Centuries of Pueblo Indian History*, Clear Light Publishers, Santa Fe.

Sekula, A. (1990), "The Body and the Archive," in R. Bolton (ed.), *The Contest of Meaning: Critical Histories of Photography*, MIT Press, Cambridge, MA, pp. 343-89.

Sontag, S. (1972), *On Photography*, Farrar, Strauss and Giroux, New York.

Spencer, L.W. (1934), "Highlights of a Trip Through 'Amerindia'," *New Mexico Magazine*, Vol. 12, pp. 7-9 and 46-51.

Spicer, E.H. (1962), *Cycles of Conquest: The Impact of Spain, Mexico, and the United States on the Indians of the Southwest, 1533-1960*, The University of Arizona Press, Tucson.

Street, B. (1992), "British Popular Anthropology: Exhibiting and Photographing the Other," in E. Edwards (ed.), *Anthropology and Photography*, Yale University Press, New Haven, pp. 122-131.

Trimble, S. (1993), *The People, Indians of the Southwest: Words and Photographs*, School of American Research, Santa Fe, NM.

Thomas, D.H. (1978), *Southwestern Indian Detours*, Hunter Publishing, Phoenix.

Urry, J. (1990), *The Tourist Gaze: Leisure and Travel in Contemporary Society*, Sage Publications, London.

Wagner, V. (1977), "Out of Time and Place – Mass Tourism and Charter Trips," *Ethnos,* Vol. 42, pp. 38-52.

Weigle, M. (1988), *The Lore of New Mexico*, University of New Mexico Press, Albuquerque.

Weigle, M. (1989), "From Desert to Disney World: The Santa Fe Railway and the Fred Harvey Company Display The Indian Southwest," *Journal of Anthropological Research*, Vol. 45, pp. 115-137.

Weigle, M. and Babcock, B. (eds) (1996), *The Great Southwest of the Fred Harvey Company and the Santa Fe Railway*, Heard Museum, Phoenix, AZ.

Younger, E. (1983), "Changing Images: A Century of Photography on the Hopi Reservation," in E. Younger and V. Masayesva Jr. (eds) *Hopi Photographers/ Hopi Images*, Sun Tracks/University of Arizona Press, Tucson, pp. 13-40.

Younger, E. and Masayesva, V. Jr. (eds) (1983), *Hopi Photographers/Hopi Images*, Sun Tracks/University of Arizona Press, Tucson.

Acknowledgements

We were grateful to receive the Graduate Research Partnership Program Summer Research Grant at the University of Minnesota which provided funding to conduct this research. We also wish to give thanks to the Governors and staff of the Eight Northern Indian Pueblos Council, Inc. for permission to use photos in the publication. Also, much appreciation goes to former *Visitors' Guide* coordinators Jesse Davila (San Ildefonso Pueblo), Walter BigBee (Comanche) and Theresa True (San Juan Pueblo/Navajo), for sharing their insights and access to archival information – Kúdaawóháa.

Chapter 3
Ancient Greek Theatres as Visual Images of Greekness

Vassiliki Lalioti

Introduction

In this chapter I investigate the ways in which ancient Greek theatres, as material monuments of a glorious culture of the past and as emblems of modern Greek ethnic identity, are re-presented in brochures published by the Greek National Organization of Tourism (GNOT), in order to attract tourists. Modern Greeks externalize aspects of their ethnic identity, and by using these photographs for touristic reasons they understand and define themselves in an international context. The way that the space – ancient monuments and landscape – has been photographed creates the sense of stability and un-changeability of the ancient Greek culture and reinforces its de-historicization. These photographs call (mainly European) tourists to come to Greece to 'see' the 'authentic' Greek culture, as if no time has intervened from classical antiquity until today. Although the 'meaning' that can be attributed to these photographs through a visual analysis may be theoretically open-ended, it is also historically and culturally determined. Thus, within the wider context of power relations that exist between Western Europe and its periphery (to which the familiar exotic Greece belongs), the Hellenic state policy on tourism focuses on the exchange of sun and culture capital for economic capital and national profit. These photographs are analysed as self-representations, as visual images of the uniqueness of Greekness that reproduce the local-global dimension of the ancient heritage, a 'weapon' that Greeks have in negotiating with more powerful 'others'.

Contemporary anthropological research on tourism and photography recognizes their central place in organizing social interactions between different groups, and in negotiating images of 'us' in relation to the 'others'. Photographs are analysed as socially constructed artifacts that reveal aspects of the culture depicted as well as the culture of the person who takes the photograph (Edwards 1992; Scherer 1992; Lutz and Collins 1994; Collier 1995; Mead 1995). Recent discussions 'concentrate more on the social contexts of making and using images and less on the photography as text' (Ruby 1996: 1346), thus giving emphasis to the process of production and negotiation of their meaning, which is not conceived to be something fixed and final. Since one of anthropology's main concerns is the investigation of identities and othernesses, it may contribute significantly to the study of tourism, a significant meeting point of 'us' with the 'others' (Graburn 1989; Bruner 1991; Galani-Moutafi 1995; Abram and Waldren 1997).

Literature on the role of photography in tourism is mainly concerned with the tourists' practice of photographing local people or sights at tourist destinations as well as with the relationship between photography and the activity of traveling itself. Many studies focus on the interaction between tourists and the environment (Urry 1995; Markwell 1997, Crawshaw and Urry 1997) and on revealing the socially constructed 'nature' of seeing (Berger 1972; Barthes 1981; Osborne 2000; Urry 2002). In Greece tourism has mainly been the object of sociological and anthropological studies investigating its effects on the socioeconomic development of the country in the last fifty years (Tsartas 1989) and the transition in economic orientation from agriculture to tourism (Scott 1985; Galani-Moutafi 1993, 1994). Although tourism has often been identified with the practice of photographing itself, ethnographic accounts of tourism have rather neglected its visual aspects. In this chapter I intent to analyse the photographs of ancient Greek theatres that are included in brochures published by the GNOT. Since these photographs have been taken by Greeks and aimed at attracting foreign tourists, they are analysed as collective self-representations which are thus subject to the influences of their social, cultural, and historical contexts of production and consumption.

Although these photographs are not anthropological, in the sense that they were not taken by an anthropologist in order to use them as ethnographic data, they are used as such because meaningful information can be derived by them in the study of the self-representation of Greeks to the 'Other' (mainly European tourists). Current visual anthropological literature treats photographs as 'cultural artifacts', which give meaning to 'political, economic, and social understandings, preconceptions and stereotypes' (Scherer 1992: 33). Within this frame, the analysis of the photographical material collected for this study contributes to a better understanding of the process of negotiation of ethnic identity in the field of tourism, which 'provides the setting for people to reconsider how they identify themselves, and how they relate to the rest of the world' (Abram and Waldren 1997: 10). Additional material (other photographs, texts, interviews, newspapers) supports the 'reading of the image' (Edwards 1992) since it helps to place these fragments in the social context of their production and consumption.

These photographs are images of material monuments of classical antiquity, which not only belongs to the Greek ethnic past, but is also perceived to be the origin of the civilization of Europe. Since tourism constitutes a 'meeting point of the local with the global, a sphere where the local directly experiences the local-global relationships' (Yalouri 2001: 128), these photographs can be analysed as cultural objects that reveal wider ideas and perceptions of the local/global dimension of the Greek ethnic identity.

As representations of what are considered to be generic markers of 'Greekness' made by Greeks, these photographs are self-representations, which reflect and reveal relations of power and cultural/political hegemony within the European context. Power relations are always present in representations of the self and the other and, according to Galani-Moutafi, mass tourism may be seen 'as a new kind of mass invasion of the western world to the countries of the periphery' (1995: 38). Thus, while the technologically developed countries promote this development as a tourist attraction, they expect that the 'traditional' and 'authentic' character of the periphery will be preserved. Photographs, like the language that is used in the tourist brochures of the GNOT, tend to represent the Greek ethnic self in an eternal present (Fabian 1983), whose position has been predetermined by a dominant discourse in Greece itself and in the Western centre. Although brochures promise the tourists that they will be transformed during their trip to the familiar 'exotic' periphery of Europe, to which Greece belongs, the locals remain unaffected by the changes that take place around them. During the process of self definition, members of ethnic groups select elements of their past that are appropriate and ignore others, according to the conditions and the needs of their present. The past functions as a means for authentification of current Greek culture, one which is perceived to be the survival of an ancient, significant one, whilst remaining stable and unchanged through centuries, thus making Greece an attractive tourist destination. These images are representative of the way modern Greek culture is perceived by Europeans but also by the Greeks themselves, who have accepted these interpretations and have transformed them to cultural identities (Htouris 1995: 50).

Ancient Greek Theatres: Culture versus Tourism

Ancient Greek theatres were built and used for the staging of ancient Greek drama performances (tragedy and comedy). The initial forms of the theatrical space are dated back to the sixth century B.C., although theatres continued to develop until the Hellenistic times (323-31 B.C.). Today, ruins of 45 ancient theatres can be found in Greece (Athens, Peloponnese, Aegean islands, Crete, etc.), 12 in Asia Minor, and 11 in Sicily and Southern Italy although only a very small number of them are well preserved (Stefosi and Kostopoulos 1996). Restoration and conservation – which usually meant rapid reconditioning – of these theatres in modern Greece, usually coincided with the few art festivals which became established during the 1950s and 1960s. Tensions related to the essential protection of the theatres between the archaeological services, under the authority of which these ancient monuments lie, and the tourism services, are very common. Extensive use of specific theatres, either for visiting or for theatrical performances is considered to have accelerated their physical deterioration (Boletis 1998; Spathari 1998).

Festivals of ancient Greek drama, which take place in well preserved ancient theatres (Epidaurus, Dodona, Filippi etc) constitute a permanent feature of Greek cultural activity. Placed among the most powerful means by which the concept of the Greek cultural superiority against the rest of the world is constructed, ancient drama festivals reinforce the distinctive Greek identity (Lalioti 2002a), and function as tourist attractions, which bring more money to the country. Official statements made by politicians, often refer to various ancient drama festivals as economic resources:

> The ancient theaters' revival will most of all enrich the content of our contemporary life, but at the same time will give a monopolistic character to Greek tourism, providing its right social dimension. Because especially in Greece, tourism and culture go together … We must, therefore, act in awareness of the fact that the goods of the Greek civilization consist a universal inheritance, and that our monuments are diachronic … (N. Sifounakis, Minister of Culture, *To Vima*, 23 July 1995).[1]

During the past two decades the Festival of Epidaurus, the festival most promoted in both the domestic and the international markets, was under the supervision and the administration of the Greek National Organization of Tourism. In 1999 these responsibilities passed to the state company 'GNOT – Greek Festival SA', which took over all the GNOT cultural activities. The theatre of Epidaurus is exclusively used for the staging of ancient Greek drama performances since 1955, and no other cultural activities are allowed there (Lalioti 2002b). Non-Greek theatrical groups were not allowed in the theatre until 1994 and since then one performance by such a group has been established: 'allowing them to show their work' (President of GNOT, author interview, 1998). The aim of this 'concession' is not just to promote the discussion on ancient drama through the presentation of different views, but to promote Greek culture abroad: ' … imagine what they say when they go back to their country … It is a promotion of our culture' (actress with long experience in ancient drama, author interview).

The issue of preservation of ancient monuments reflects the wider debate that takes place over the museum status of important monuments of a culture of the past, or their active incorporation into contemporary life. The tendency to reuse the monuments in Greece was combined with the touristic development that took place during the 1960s and 1970s without the necessary scientific and technical support. Even today, the grounds on which the Archeological Resources Fund decides on the preservation and promotion of ancient theatres and castles around the country is that this initiative 'contributes to the cultural battery of the country' and corresponds to the quest of the E.U. for the 'parallel development of culture and tourism' (*Eleftherotypia*, 17 July 1996). The key issue of the debates between agents of tourism and culture in reference to ancient theatres in Greece is whether the Festival of Epidaurus should be under the control of the Ministry of Tourism or if it should return to 'its natural environment' that is, the Ministry of Culture. It reflects the power relations that exist in the country and constitutes an essential aspect of the national policy concerning the role of the past in the creation of the ethnic image in the local as well as in the international contexts. The debate is far from concluded and one

---

1 Translated by the author from Greek.

can see other refracted issues in it such as culture (protector) versus 'commerce' (tourism); localism versus globalization.

The Photographs

According to the World Tourism Organization, Greece occupies the fifteenth place in terms of number of arrivals during the year 2000 (13,095,545 foreign tourist arrivals). The majority, 70.4 per cent, of foreign tourists came from the EU, 21 per cent from other European countries, and only 2.5 per cent from the US (GNOT 2002: 6-7). The 'one-dimensional touristic product sun-sea' and the fact that 'the comparative advantages of the Hellenic tourism are mainly inherited (natural environment and cultural heritage)' are two of the main characteristics of the Greek tourism (GNOT 2002: 1). Brochures published by the GNOT, that any tourist can find in the offices of the organization around the country today, contain photographs of various kinds of attractions: antiquities, natural environment, modern monuments, activities, maps, texts with historical information about places worth visiting, and general useful information (hotels, telephone numbers, transportation, accommodation, etc). Although I did not conduct any statistical analysis, it would not be too risky to argue that photographs of antiquities and natural beauties of Greece make up the majority of the photographs included in these brochures, visualizing thus the dipole sun-antiquities promoted by the GNOT policy for attracting foreign tourists. Focusing on the photographs of ancient theatres one can easily ascertain that they occupy a privileged place in brochures of areas widely known for their well-preserved ancient theatres.

What struck me from the first moment I looked at these photographs was the obvious similarity of these images: vivid colours, the stones of the theatre, beautiful view, the natural environment that surrounds the theatres (sun, mountains, blue sky, and trees), strong daylight. And in almost all of them there are no people.On the cover page of the Brochure of Peloponnese, for example, we can see five photographs of what are considered to be the most highly valued touristic attractions of the area: ruins from ancient Olympia, the Lion's Gate in Mycenae, the castle in Methoni, the ancient theatre of Epidaurus, and a young girl sitting in a table with a bottle and a glass of (most probably) retsina, bread, and a plate of Greek salad. In the blue blurred background we can see the sea. The light clothes of the girl, the light wind in her hair, and the strong daylight in all photographs suggest it is summer. Inside the brochure, there is a big picture of the theatre of Epidaurus (see Figure 3.1): here, like in the photograph of the cover, we can see the theatre located in a beautiful landscape (green hills around, blue sky), strong daylight and a human figure in the down left corner serving as an index of the size of the monument. The other two photographs on the page show the nice port of Porto Heli and the Palamidi Fortress in Nafplio (a historic city in the Peloponnese). In the text referring to the site of Epidaurus we read:

On a hillside, within the sanctuary, lies the theatre of Epidaurus (third century BC), the most famous and best preserved of all ancient theatres in Greece. Built of limestone, it can seat 12,000 spectators. Every summer it comes alive. Attending a performance of ancient drama in this theatre is almost a mythical experience. Never to be forgotten. A catharsis of the soul. (GNOT 1998: 4)

In the same brochure we can see a photograph of the ancient theatre of Megalopolis. It is not as impressive as the theatre of Epidaurus, but one can see the strong daylight and elements of the natural environment surrounding what has remained of the ancient monument. In the opposite page we read:

> ... whoever roams through Arcadia [the prefecture] today will get to know a part of the country that has remained virtually unchanged since antiquity. Arcadia is a natural wonderland whose mountain dwellers have preserved it intact up to today Photographs above show a sample of the local architectural tradition of the town of Dimitsana and the Byzantine Holy Monastery of Philosophos. (ibid: 6)

In the cover page of the brochure of Delphi, 'home of the sanctuary and oracle of the god of light Apollo [which] was believed to be the "navel of the world" by the ancient Greeks ... the largest religious and spiritual center in Hellas ...' (ibid: 1), we can see a picture of the *Charioteer* (478 or 474 B.C.) and the entrance to the temple of Apollo (fourth century B.C.). The photograph of the ancient theatre of the area is a big picture covering almost two pages. The monument is well preserved, the natural environment is impressive, and tourists have come to visit it during summer (this is suggested by the strong daylight and light clothes of the people). Other photographs in the brochure show finds exhibited in the local museum and 'other places of interest in the region': the harbour at Itea, a small village in Amphissa, the picturesque Galaxidi, the ski resort in Parnassos mountain, a mosaic from the Byzantine Monastery Hossios Loukas (eleventh century).

The brochure of Macedonia, like the ones mentioned above, contains photographs of a) antiquities: (in the cover page) a detail of the cuirass of Philip II, the head of Alexander the Great, (inside) the Byzantine walls of the city of Thessaloniki, detail from the arch of Galerius, the gold larnax found in the royal tomb in Vergina, the archaeological site of Dion, Byzantine frescoes from churches in Kastoria, etc., and, b) aspects from modern life: views of the big cities, c) natural beauties of the area (beaches, mountains), d) local architecture and e) elements of folk culture (traditional costumes from the area and traditional carnival disguises). On the first inside page there is a brief presentation of the history of the area entitled 'Macedonia: 4,000 years of Greek History and Civilization' concluding: 'In 1912, Macedonia was liberated and incorporated to the Greek territory. Since then, it has been part of the Greek state, a symbol of the origin of the Greek spirit and civilization' (GNOT 2002b:1). The photograph of the ancient theatre of Filippi (near the city of Kavala) depicts the monument surrounded by trees. It coexists with photographs of other antiquities (the archaeological site of Thassos), beautiful beaches (the campsite at Kavala) and elements characteristic of the 'local colour' (architecture, small boats in the picturesque ports of Kavala and Thassos).

Images of Greekness

According to Urry, photography seems to transcribe reality in pieces and the 'images produced appear to be not statements about the world but pieces of it, or even miniature slices of reality' (2002: 127). All photographs are as much the result of an active signifying practice in which those taking the photo select, structure, and shape what is going to be taken. Every group selects certain locales and activities which perceive them to be representative of the public image, areas and structures which they expect the stranger to recognize: images 'that are the pride image of a community' (Collier 1995: 240). Ancient theatres are essential components and easily recognizable monuments of the ancient Greek culture. Photographs of the ancient theatres coexisting with those of the Byzantine and the modern era make the Greek homeland as much a state of mind as a place on the map (Kouvertaris 1987). All photographs together visualize moments of the long history of the Greek culture, which began in prehistoric times, continued in classical antiquity, and through the Byzantine era came to our days. A visual narration of continuity of the Greek ethnos is thus created, which constitutes one of the main criteria of reference and knowledge of the self and the other in Greece. 'The determination of the modern Greek culture in relation to the ancient heritage, the Byzantine tradition … and the European modernity are the context within which … the current Greek cultural identity is been produced' (Htouris 1995: 54).

The accumulation of documentary evidence functions towards the same direction: ethnic groups confirm an apparent continuity that is recorded while it is evident that it has been lost from memory. Continuity is seen as the result of a secular, serial time, and because the experience of continuity has been forgotten, nations need narratives of identity (Anderson 1991). Photographs of ancient theatres as archaeological sites are used as witnesses of evidential things, which must be remembered as 'our own Documents become narrations of something lost (we do not remember it anymore) and at the same time they create a conception of personhood, of identity while they are both set in homogenous, empty time.

The tourism industry uses photography to sell tourism products through brochures, guidebooks or postcards. It selects photographic images based on two criteria: subject and visual quality, reducing a complex place to a few iconic images, which present only a partial picture, actively distort the identity, trivialize the place, and deny the process of change (Urry 2002). In the photographs described above, Greece becomes its ancient theatres, which stand in the same place for centuries surrounded by the same natural environment. The absence of human forms emphasizes the sense of timelessness that these photographs intend to create, and de-historicizes the ancient Greek culture. Everyday discourse supports these poetics of achronicity and unchangeability: 'Today, most of these ancient theatres are silent remnants. Yet, they do not cease to "speak" through their stones; they are witnesses of the unique civilization of Greeks ... the landscapes in which we live and are the same with the landscapes of our ancestors...' (Stelios, interview by the author).

Although the material environment that surrounds us also evolves, it gives the impression that it always remains the same, creating thus a feeling of stability and permanence (Bender 1993) for the Greek ethnic group. When regarded as material places, ancient theatres are not simply historical monuments; they are conceptualized as history itself and become 'tangible history', creating a distinctive form of space: the physical environment of the eternal, ideal Greece (Yalouri 2001). These photographs, thus, propose a travel through space, which is also a travel through time. Space (natural and built) is reconstructed in a way that conveys a sense of stability and continuity from antiquity until nowadays that history does not support – the natural environment has evolved and the Greek culture has changed. By promoting clearly pictured images of the ancient theatres in tourist brochures, Greeks externalize perceptions of their national identity, and at the same time they reproduce or transform the ways they understand and define themselves in an international context. The elevation of the natural environment as an agent and instigator of culture in modern Greece has been widely discussed in relation to ethnic identity issues (Tziovas 1989). Especially the notion of 'light' (ancient theatres are bathed in clear and strong sunlight in all photographs) far from being only something physical it might also refer to the 'light' of civilization that ancient Greeks gave to the rest of the world (or at least of the western world).

The narrative of de-historicization of the Greek culture is further supported by the images created by the texts that accompany the photographs in the brochures: 'On this "flaming red Argive earth" celebrated by the poet, "where the poppy flames still brighter", you'll hear the most sublime voices of the Greek land – Homer, Aeschylus, Sophocles' [about Argolida: the prefecture where the theatre of Epidaurus is located, Brochure of Peloponnese] (GNOT 1998: 4). The ancient poets can still be heard today. The present and the past are perceived to coexist in 'homogenous, empty time' (Lekkas 1996: 211) and space, the ethnic time and space. Like all ethnic groups, Greeks regard themselves as an entity that is 'essentially immovable and a-historic, almost transcendental' (ibid.). The Greek ethnos is represented in an eternal present and its position is predetermined by a dominant discourse cultivated by Greeks themselves as well as by Europeans (another transcendental, homogenous entity), whom they want to attract. Although these narrations 'may seem to be simplistic overestimations' they actually 'organize and give meaning to the touristic interactions for both, the tourists and the locals' (Galani-Moutafi 1995: 35).

These pictures produce an imaginary space and time supported by the 'objectivity' of the photograph. Their creators, however, intentionally promote a specific aspect of the 'reality' in order to satisfy the expectations and the already existent images in the mind and the consciousness of foreign visitors.

The photographic images of ancient Greek theatres evoke to those who see them particular associations 'in a powerful synecdochal and iconic way', since they can 'conjure up an entire site ... by representing only a fragment' (Crouch and Lübbren 2003: 5). Ancient theatres become symbols of the classical antiquity and Greece a touristic site where the ruins of the 'authentic' Greek culture, which also constitutes the origins of the European civilization, can be seen. The creation of these images actually presupposes the models of understanding and the horizon of the historical and social consciousness within which they are produced. The bipolar imagery of nature–antiquities nowadays continues the long tradition of the European aristocracy whose visit to the Italian and Greek archaeological sites constituted a prerequisite for the completion of the Grand Tour, the participants' baptism to the culture of antiquity (Kardasis 2000).

'Us' and 'Europe'

Traveling to Greece

Long before Greek Independence (1832), English, French, and German travelers made the long and dangerous trip to Greece stimulated by their passionate belief in the idea of Hellas. A feeling of nostalgia motivated these trips as 'visiting the land itself, recovering the sites and the works of art, enhances the sense of loss, in that one sees more clearly what once was' (Constantine 1984: 4). Eighteenth century Europe created, through its admiration and longing for the values ascribed to ancient Greece, its own myth and gave Hellas the dimensions of an ideal. Europe interpreted ancient Greece according to the needs of the construction of a common western identity, making thus the 'control of the image of the Greek past equally essential for Europeans' (Alexandri 1997: 97). It was part of the wider quest for the genesis of nations that constituted the base of a hierarchically developed civilization. In 1670, Athens became a holy city and the birthplace of the now secular European civilization. 'The European continent will, from now on, become tantamount to the place of civilization, to use the term which the eighteenth century devised, in order to, among other things, distinguish Europe from other continents, from wildness and barbarism' (Giakovaki 1997: 80), when Europe discovers the 'other'.

Visits to the land of their ideals led European travelers to compare the remnants of a civilization of the past with the way of life of the contemporary inhabitants. In their eyes, the comparison was revealing: there was nothing left from that glorious past, and what they brought back was a 'tension, ideally a creative one, between facts and the ideal, scholarship and imagination' (Constantine 1984: 211). What European travelers imagined to have found was a loss, which although a historical one, they perceived as a synchronic lack of culture. In both cases, either in treating Greeks as living ancestors or as uncultured Orientals, they denied them having a culture.

During the period of genesis of the modern Greek nation-state (nineteenth century) Greeks used this ambiguous picture to achieve their aim. It was very convenient for Greeks to accept 'the passive role of the living ancestors' (Herzfeld 1998: 141), although it later led to criticisms (inside and outside Greece) for their inability to be worthy of that past, and equal to the rest of Europe. Dependence on the West made Greeks cultivate feelings of estrangement inside their own country, which coexisted with a consciousness of the historical and cultural centrality of Greece to European civilization. What is peculiar to Greece's case is that everyone seems to have appropriated its lost cultural hegemony. And the problem for Greek nationalism was to distinguish an 'us' from 'them' while 'the very measure of your fame is the degree to which your definitions of self have been appropriated by others. How, in short, are you to play the role of being exclusively universal?' (Just 1995: 290).

Elements of a Greek Ethnic Identity

The photographs of ancient Greek theatres that are included in the brochures of the GNOT are the results of a process of selection intending to create a specific image of modern Greece as a unique touristic destination. This image reflects elements of a specific Greek ethnic identity construction process, which is also characterized by some sort of selectivity. By taking photographs of ancient theatres for tourist purposes, Greeks select and appropriate images of their ethnic self that satisfy current needs: economic profit and an equal place amongst the powerful of the West. By giving 'photographic emphasis' on particular elements of their culture (sun-antiquities) Greeks create a specific narrative of what they consider to be generic markers of Greekness.

The ethnic 'self' in Greece is perceived through a specific feeling of superiority, which rests initially on the achievements of a culture of the past and consequently on contemporary image justified by the fact that Greeks are 'the distinctive heirs of the ancient Greek civilization and ... that ancient Greece constitutes the perpetual cultural matrix of Europe' (Veikos 1993: 38). On the other hand, the image of the East is always present, representing and justifying usually the negative aspects of the Greek character and the contemporary inability to keep up with progress in Europe. Both these poles are present in everyday discourse and express a contradiction, which exists throughout the whole spectrum of the codes through which the collective self-presentation and self-cognition are counterbalanced (Herzfeld 1987).

Contradictions, however, are not a unique Greek characteristic but a reality that is common to all ethnic identities. In the case of Greece, however, they represent 'the counterbalance of self-knowledge and self-presentation against more powerful others' (Papataxiarhis 1998: xix). They constitute neither a restriction, nor anything negative in the process of construction of an ethnic identity, but rather they create relationships of multileveled negotiation of meaning. For Europeans, Greece has been a place of 'their own' although far away from them in time (ancestral and thus removed through mythic time) and in space, but not 'them'. In the same way, Greeks perceive of themselves as belonging to Europe but as not being Europeans in every sense.

There is a way, however, in which Greeks are supposed to be different, as far as the way they handle their sense of superiority is concerned. The reconstitution of homogenous identities largely through appeal to an ancient ideal is not a unique phenomenon, but it is only in Greece that the revival of an ancient ideal is supported by the agreement of almost the entire world. So ready are others to accept the idea of Greece as ancient that Greeks themselves at times find it difficult to insist on their modernity. But for Greeks 'to be modern is to discover tradition, to exoticize it Greece, as a country in the margins of Europe, is characterized by two diverging forces: 'the certainties of [its] past (often manufactured by the State and its elites, often in association with the West), and the uncertainties of [its] modern vulnerability' (Sant Cassia 2000: 298).

For some Greek scholars (see for example Tsaousis 1998), Greece is a country experiencing a crisis of identity because the content of this identity remains obscure and ambiguous. According to this view, 'the identity crisis is the central problem of the neo-hellenic society, the constituent element of the contemporary Hellenism, and the axis around which our modern history revolves' (ibid: 17). Thus, there is a rather defensive definition of Hellenism, which seeks to establish identity not on the affirmation of 'us', but on the rejection of the 'other', leading to an effort to retain its integrity through isolation and ethnocentrism. There is also a tendency for many non-Greek academics to see Greece as a country constantly in pursuit of its identity. This approach could reflect western rationalistic views of identity as something homogenous and unified, while notions of conflict and competition are also present. It reproduces the notion of the marginality of Greece that exists in international affairs and is reflected in 'the very marginality of Greek ethnography in the development of anthropological theory, where ... the Greeks of today are a people neither dramatically exotic nor yet unambiguously European' (Herzfeld 1987: 20).

Seen from the point of view of cultural hegemony, Greece is a culturally dominated society (Argyrou 1996) and 'Europe' has always been a crucial factor in the process of constructing modern Greek ethnic identity. It has been present in the form of the 'other' with whom Greeks either identify or differentiate themselves from. Moreover, Europe, on a more pragmatic, political level, has been directly and indirectly involved in the formation of the contemporary Greek nation state and its survival in the world system, thus effecting further constructions, interpretations and reflections on the ethnic self (Politis 1997; Kremmydas 1997). The West and Europe is a potent, even dominating symbol in Greece. In the last century it was used to tie all Greeks to a particular identity and at the same time subjugate it to other, more powerful societies.

Figure 3.1   The ancient theatre of Epidaurus

Conclusions

My aim in this chapter was to investigate the way ancient Greek theatres are depicted in photographs included in promotional material (brochures) published by the Greek National Organization of Tourism in order to attract foreign (mainly European) tourists. Since tourism consists a significant meeting point of 'us' with the 'others', of the local with the global, meaningful information is derived from these photographs as far as the study of collective self-representations of Greeks, as members of an ethnic group, is concerned. The analysis of the photographs was positioned within two contexts: a) perceptions of 'Us' and the 'Other', and b) the power relations that exist between Greece (the periphery) and Western Europe (the hegemonic centre).

The striking similarity of these photographs, whose creators have selected a specific way of depicting the space – ancient theatres and natural environment but no people in them – creates a narrative of continuity and unchangeability of the ancient Greek culture until today, as if no time has intervened. Hellenism is perceived to be diachronic, continuous, unified and homogenous and to survive within empty time inside material monuments of a glorious culture of the past and in the natural environment. The texts and the rest of the photographic material included in these brochures support this de-historicization of the Greek culture, promoting the bipolar images of nature-antiquities, two elements that make Greece a unique touristic destination for Europeans.

In the manner that taking photographs is a selection process, ethnic identity construction is a selection process too. By taking photographs of ancient theatres for the purpose of tourism promotion, Greeks select and appropriate images of their ethnic self that satisfy current needs. Although marginal to the wider European context of power relations, Greece counterbalances its ambiguous position (modern or traditional–exotic) by reproducing the local-global dimension of its ancient heritage and by exchanging sun and culture capital for economic capital and national profit.

References

Primary Sources

Eleftherotypia (Greek daily newspaper), 17 July 1996.
Greek National Organization of Tourism (1998), Brochure of Peloponnese Greece, 6, Directorate of Market Research and Advertisement Publication and Audio Visual Means Department.
Greek National Organization of Tourism (2002a), Brochure of Delphi Greece 1, Directorate of Market Research and Advertisement Publication and Audio Visual Means Department.
Greek National Organization of Tourism (2002b), Brochure of Macedonia Greece, 6, Directorate of Market Research and Advertisement Publication and Audio Visual Means Department.
Greek National Organization of Tourism (2002c), Tourist Policy 2002-2006, Aims and Perspectives, accessed online at <http://www.eot.gr/3/03/gc30000.html>, 15 January 2003.
Sifounakis N. (1995), 'Interview', To Vima (Greek daily newspaper), 23 July 1995.

Secondary Sources

Abram, S. and Waldren, J. (1997), 'Introduction: Tourists and Tourism-Identifying with People and Places', in S. Abram, J. Waldren and D.V.L. Macleod (eds), *Tourists and Tourism*. Berg, Oxford-New York.
Alexandri, A. (1997), 'The Greeks and Us', *Contemporary Issues*, No. 64, pp. 97-104.
Anderson, B. (1983), *Imagined Communities*, London, Verso.
Argyrou, V. (1996), *Tradition and Modernity in the Mediterranean: The Wedding as Symbolic Struggle*, Cambridge University Press, Cambridge.
Barthes, R. (1981), *Camera Ludica*, Hill & Wang, New York.
Bender, B. (ed.) (1993), *Landscape: Politics and Perspectives*, Berg, Providence-Oxford.
Berger, J. (1972), *Ways of Escape*, Penguin, Harmondsworth.

Boletis, K. (1998), 'The Conservation and Use of Ancient Theatres in Modern Greece – Current Issues Relating to the Epidauros Theatre', *Journal of Mediterranean Studies*, Vol. 8, No. 1, pp. 14-19.

Bruner, E. (1991), 'Transformation of Self in Tourism', *Annals of Tourism Research*, Vol. 18, No. 2, pp. 238-250.

Collier, J.J.R. (1995), 'Photography and Visual Anthropology', in P. Hockings (ed.), *Principles of Visual Anthropology*, Mouton de Gruyter, Berlin and New York, pp. 235-254.

Constantine, D. (1984), *Early Greek Travelers and the Hellenic Ideal*, Cambridge University Press, Cambridge.

Crawshaw, C. and Urry, J. (1997), 'Tourism and the Photographic Eye', in C. Rojek and J. Urry (eds) *Touring Cultures*, Routledge, London, pp. 176-195.

Crouch, D. and Lübbren, N. (2003), *Visual Culture and Tourism*, Berg, Oxford and New York.

Edwards, E. (1992), 'Introduction', in her (ed.) *Anthropology and Photography: 1860-1920*, Yale University Press, New Haven and London, pp. 3-17.

Fabian, J. (1983), *Time and the Other: How Anthropology Makes its Object*, Columbia University Press, New York.

Galani-Moutafi, V. (1993), 'From Agriculture to Tourism: Property, Labor, Gender and Kinship in a Greek Island Village', *Journal of Modern Greek Studies*, Vol. 11, pp. 241-270.

Galani-Moutafi, V. (1994), 'From Agriculture to Tourism: Property, Labor, Gender and Kinship in a Greek Island Village', *Journal of Modern Greek Studies*, Vol. 12, pp. 113-131.

Galani-Moutafi, V. (1995), 'Tourism Perspectives: Invention and Authenticity', *Contemporary Issues*, Vol. 55, pp. 28-39.

Giakovaki, N. (1997), 'Locating Ancient Greece: The Europeans and the Emergence of a New Country in Modern Times', *Contemporary Issues*, No. 64, pp. 76-80.

Graburn, N. (1989), 'Tourism: The Sacred Journey', in V. Smith (ed.), *Hosts and Guests: The Anthropology of Tourism*, University of Pennsylvania Press, Philadelphia, pp. 21-36.

Hertzfeld, M. (1987), *Anthropology Through the Looking Glass, Critical Ethnography in the Margins of Europe*, Cambridge University Press, Cambridge.

Htouris, S.N. (1995), 'Culture and Tourism: Tourism as Network of Experience Production', *Contemporary Issues*, Vol. 55, pp. 48-56.

Just, R. (1995), 'Cultural Certainties and Private Doubts', in W. James (ed.), *The Pursuit of Certainty*, Routledge, New York and London, pp. 287-302.

Kardasis, V. (2000), *Hellas: A Journey Through Time*, Greek National Organisation of Tourism, Athens.

Kouvertaris, Y. (1987), 'People, Identity, and Land', in Y. Kouvertaris and B.A. Dobartz (eds), *A Profile of Modern Greece in Search of Identity*, Clarendon Press, Oxford, pp. 1-15.

Kremmydas, V. (1997), 'Perception and Uses of Antiquity in the Greek Enlightenment', *The Citizen*, Vol. 23, pp. 24-29.

Lalioti, V. (2002a), 'Social Memory and Ethnic Identity: Ancient Greek Drama Performances as Commemorative Ceremonies', *History and Anthropology*, Vol. 13, No. 2, pp. 113-137.

Lalioti, V. (2002b), 'Social Memory and Ancient Greek Theatres', *History and Anthropology*, Vol. 13, No. 3, pp. 147-157.

Lekkas, E.P. (1996), *The Nationalistic Ideology: Five Working Hypotheses in Historical Sociology*, Katarti, Athens.

Lutz, C. and Collins, J. (1994), 'The Photograph as an Intersection of Gazes, The Example of National Geographic', in L. Taylor (ed.), *Visualizing Theory: Selected Essays from V.A.R. 1990-1994*, Routledge, New York and London, pp. 363-384.

Markwell, K. (1997), 'Dimensions of a Nature-based Tour', *Annals of Tourism Research*, Vol. 24, No. 1, pp. 131-155.

Mead, M. (1995), 'Visual Anthropology in a Discipline of Words', in P. Hockings (ed.), *Principles of Visual Anthropology*, Mouton de Gruyter, Berlin-NewYork, pp. 3-10.

Osborne, P. (2000), *Traveling Light. Photography, Travel and Visual Culture*, Manchester University Press, Manchester.

Papataxiarchis, E. (1998), 'Ethnography and Self-Knowledge' in M. Hertzfeld, *Anthropology Through the Looking-Glass*, Alexandria, Athens, pp. xvi-xxvi.

Politis, A. (1997), 'From the Roman Emperors to the Glorious Ancestors', *O Politis*, Vol. 32, pp. 12-20.

Ruby, J. (1996), 'Visual Anthropology', in D. Levinson and M. Ember (eds), *Encyclopedia of Cultural Anthropology*, Vol. 4, Henry Holt and Company, New York, pp. 1345-1351.

Sant Cassia, P. (2000), 'Exoticizing Discoveries and Extraordinary Experiences: Traditional Music, Modernity and Nostalgia in Malta and Other Mediterranean Societies', *Ethnomusicology*, Vol. 44, No. 2, pp. 281-301.

Scherer, J.C. (1992), 'The Photographic Document: Photographs as Primary Data in Anthropological Enquiry', in E. Edwards (ed.), *Anthropology and Photography: 1860-1920*, Yale University Press, New Haven and London, pp. 32-41.

Scott, M. (1985), 'Property, Labor and House Economy: The Transition to Tourism in Myconos, Greece', *Journal of Modern Greek Studies*, Vol. 3, No. 2, pp. 187-206.

Spathari, E. (1998), 'Ancient Greek Theatres: Their Protection and Promotion – Problems and Perspectives Concerning their Use', *Journal of Mediterranean Studies*, Vol. 8, No. 1, pp. 20-26.

Stefosi, M. and Kostopoulos, N. (eds) (1996), *Ancient Theatres*, Itamos, Athens.

Tsaousis, D.G. (1998), 'Hellenism and Greekness: The Problem of the Modern Greek Identity', in D.G. Tsaousis (ed.), *Hellenism and Greekness: Ideological and Experiential Axis of the Modern Greek Society*, Estia, Athens, pp. 15-26.

Tsartas, P. (1989), *The Social and Financial Impact of Touristic Developments on the Prefecture of Cyclades and Particularly on the Islands of Ios and Serifos in 1950-1980*, EKKE, Athens.

Tziovas, D. (1989), *The Transfigurations of Nationalism and the Ideology of Greekness in the Interwar Period*, Odysseas, Athnes.

Urry, J. (1995), *Consuming Places*, Routledge, London.

Urry, J. (2002), *The Tourist Gaze*, Thousand Oaks, London.

Veikos, T. (1993), *Εθνικισμός και Εθνική Ταυτότητα* [*Nationalism and National Identity*], Ellinika Grammata, Athens.

Yalouri, E. (2001), *The Acropolis: Global Fame, Local Claim*, Berg: Oxford and New York.

Chapter 4

The Accidental Tourist: NGOs, Photography, and the Idea of Africa

Brian Cohen and Ilyssa Manspeizer

'Nature, Mr. Allnut, is what we were put in this world to rise above.'
(The African Queen, Twentieth Century Fox Films, 1951)

Introduction

Rosie Sayer's observation, as she and her companion cruised through Central Africa, rising above all adversity, encapsulates an idea that has been firmly entrenched in Western ideology for hundreds of years. 'We' – white folk – are the civilized peoples, able and obligated to extend ourselves beyond nature. 'They' – the primitive peoples – can never completely break free from nature.

The Western conception of Africa has been informed by a long history of ideas. The vast majority of these live apart from most physical realities. They are decontextualized; and they range from idealized, patronizing, romantic, and paternalistic, to clichéd, hackneyed, simplistic, ethnocentric, and downright hateful. Agencies working in the fields of relief and development in Africa function within this cultural framework and, until recently, the imagery they have employed has been indistinguishable from that of mainstream culture.

However, towards the end of the twentieth century, many organizations began to recognize that in reproducing the prevailing Western construction of what is Africa, their promotional material has failed the very people it purports to serve. This growing awareness led, by the late 1980s, to the development of a set of guidelines through which photographers and editors could address the question of how to promote the projects of Non-Governmental Organizations (NGOs) without perpetuating demeaning ideas of what Africa is. In so doing, they were challenging generations of cultural 'knowledge'. Yet the kind of imagery that was recognized as damaging was also the most likely to raise the funds needed to support relief and development programming. Indeed, in recent years, many organizations have quietly retreated from this question for that very reason.

This chapter arises out of the experience of one of the authors working on a photographic assignment for UNICEF-Zambia in 2000. The aim of the assignment was to document some of UNICEF-Zambia's development programs. In keeping with 'progressive' trends toward the representation of aid recipients, the brief required that particular effort be made to avoid communicating stereotypical images of poverty, and to avoid endangering the welfare of those being photographed.

In this chapter, we place contemporary Western imagery of Africa in a historical and political–economic context. We describe how even well-intentioned contemporary photographs of Africa and Africans recycle old myths and reinforce degrading stereotypes that were themselves revived during colonial encounters, and strengthened with colonial photography in nineteenth century. Finally, we shall discuss the current efforts of UNICEF and other non-government organizations, or NGOs, to challenge these images, suggesting possible avenues for progress.

Seeing and Believing

Photography is a form of discourse embedded within which is the notion of 'truth'. As Francis Frith, one of the more successful nineteenth century colonial photographers, once said, 'the value of a photograph – its principal charm at least – is its infallible truthfulness' (Osborne 2000: 21-22). Frith believed – or recognized – that the power of photography resides in its apparent objectivity, fidelity to nature, and ability to secure meaning. Even though by the 1840s, when the first retouching techniques made apparent 'that the camera could lie', so compelling was the immediacy of the photograph, that photography was and still is afforded a privileged place as a location of 'truth'. It is an integral part of our system of knowledge – as Sontag (1979: 93) says, 'an instrument for knowing things'.

Despite everything we now know about photography – from nineteenth-century retouching, to twenty first-century digital technology, to questions of style, perspective, framing, composition, technique, editing, and so on – there is still something deeply convincing about the photographic image (Osborne 2000: 21-22; Ryan 1995: 54-55). When Alain de Botton (2002: 5-9), no doubt as sophisticated a user of images as any, tells us that he decided to take a break in the Caribbean after having been seduced by a photograph of a sun-drenched sandy beach in a tourist brochure, he exposes that susceptibility we all share to the power of photography. We want to believe in the image.

History, fact, reality, truth, are all proved and recycled through photographs. In a world that is now more visual than literate, photography – and subsequently television – has supplanted reality. In a sense, for an experience to become real, it must be photographed (Sontag 1979: 87).

Thus as the photograph lures us into mistaking what is real, we forget to examine the context within which the photograph was taken, distributed and used. Photographs are at best the bearers only of a partial truth, imperfect witnesses mediated by the technical processes through which they are produced. But beyond this we also understand that photographs are in essence ideological. Photographs, beyond and because of their superficial resemblance to the 'real world', convey meanings that are deeply embedded in their culture. They reinforce our view of the world; as Barthes (1977: 31) notes, they comfort, give peace of mind. Photographic images are coded images, ideological, 'value-filled meanings' existing within, rather than apart from, the 'reality' they convey. They do not exist in a vacuum (Osborne 2000; Ryan 1995; Sontag 1979; Tagg 1988: 63-64).

> A photograph is a representation whose meanings and applications derive from its position in a signifying ecology made up of other representations – other photographs, related forms of representation, critical writings, general cultural knowledge and belief and so on (Osborne 2000: 18).

Yet despite everything we know, we have learned to take photography at face value, rather than as a culturally bound means of creating and processing knowledge. We have ignored 'the historical processes and cultural frameworks through which photography, as a form of representation, acquires and communicates meanings' (Ryan 1995: 54-55).

This is what lies at the heart of the NGOs' conundrum: to produce material on Africa and its peoples is to do so from within a cultural framework that has determined, invisibly and persuasively, how Africa should be known. The contemporary Western idea of Africa, which was crystallized in the nineteenth century, (Hammond and Jablow 1977; Landau 2002) remains, not least through the medium of photography, embedded in the psyche of the so-called developed world. Colonial photography enabled its audience to see – and believe – what they already 'knew', just as contemporary imagery of Africa does for us today. Photographs confirm, they reassure. We are all – in Africa, in the West – still contending with that heritage. We still believe in the images we have created.

The Origins of the Western View of Africa

Our ideas of Africa are the product of more than five hundred years of Western imaginings. Europeans have long depicted Africa both visually and in writing, building up a history of 'knowledge' about the continent and its people. Yet when Hammond and Jablow examined more than four centuries of British writing about Africa, they found that:

> In popular writing Africa is strangely homogeneous and static; differences between past and present and between one place and another are obliterated. Africans, limited to a few stock figures, are never completely human, and Africa exhibits few changes over time (Hammond and Jablow 1977: 13).

This body of ideas, exhibiting more continuity than change, becomes what Thompson calls an 'accumulated reservoir of sense' (Thompson 1984: 12) and helps inform how we understand the images that we see today, whether they appear in *National Geographic*, coffee table books, or in NGO promotional material. Although Thompson privileges words above images, much of what he proposes can be applied to both – indeed it can be argued that the distinction between word and image is itself illusory (Tagg 1988).

However, this 'reservoir of sense' does not provide us unproblematically with a pool from which to understand images. Rather, as Williams (1973: 289) has written, 'we have to be able to explain, in related terms, both the persistence and the historicity of concepts'. Ideas may persist or reappear through time, but they are used, understood and manipulated within specific contexts and processes (see also Harvey 1996), be they historical, political or economic, in different ways. Involved in this contextualization, is the process whereby images are moved from where they were taken to where they are viewed, and are ultimately understood according to the viewer's conventions and not those of the viewed (Landau 2002: 16; Osborne 2000: 23). However, this does not mean that this reservoir of sense – which, in the way we are using it, is comprised both of ideas accumulated over hundreds of years as well as the cultural context of the viewer – represents some collectively shared understanding of how to view these images (or for that matter, understand discourse). Rather, these reservoirs represent competing and diverse values and beliefs (Landau 2002: 18; Mudimbe 1994). The challenge comes in understanding why some views are privileged over others (Thompson 1984; Wolf 1999).

Whose views are privileged is important because these reservoirs of sense are a 'creative and constitutive element of our social lives' (Thompson 1984: 5-6). In other words, the ideas held within these 'reservoirs' inform not only how we understand photographs of Africa and Africans, but also what we do with this understanding. Do we feel empathy with the people pictured, or merely pity? Would feeling empathy rather than pity lead to more charitable contributions, to increased alienation from 'Africans', or toward a greater inclination to examine structural injustices that might lead to African poverty? (Not that these are mutually exclusive outcomes). Moreover, we are not pawns at the mercy of these ideas, but actively contribute, use and change them over time, both in the production and consumption of these images.

Despite the continuity in ideas that Hammond and Jablow note, European images of Africa did change periodically, depending upon the role that Europe perceived that Africa was playing at any given moment. As the Portuguese established trading posts on the continent from the fifteenth century, written reports concentrated upon matter-of-fact descriptions of what early traders saw (Hammond and Jablow 1977: 23); visual images of Africa focused upon Europe as the center of world trade routes (Pieterse 1992: 18-20). However, as trade relations grew increasingly dependent upon the slave trade and relatively less dependent upon non-human commodities like ivory and gold, British authors began judging African behaviour rather than simply reporting it (Hammond and Jablow 1977: 23). Images from the time showed an increased alienation of Europeans from Africans (Landau 2002: 2). Africans suffered deeply at the hands of European portrayals. 'African behaviour, institutions, and character were not merely disparaged but presented as the negation of all human decencies' (Hammond and Jablow 1977:

23). Africans' humanity was questioned, and with this questioning came severe implications for Africans.

Because the British imagined Africans as savage and not quite human, proponents of the slave trade were able to justify the trade in human bondage by claiming that enslavement was actually preferable to the physical hardships and moral depravity of life for Africans in Africa, and that slavery was in fact an African's only route to salvation (Hammond and Jablow 1977: 23). Such justifications of despicable treatment for indigenous people at the hands of more powerful outsiders have been often repeated. From the mid-1850s British imperial expansion was also being touted as 'for the good of Africans' (Hammond and Jablow 1977: 91-92), just as more recently wars against dictatorial leaders have been justified according to the same criteria. However, Barrell (2000) claims that what really underlies this argument is the dehumanization of the colonial subject. It becomes acceptable to compel the 'Other' who is not quite human to become part of a colonial empire simply because they *are* not quite human.

Early colonizing efforts did not improve British attitudes towards Africans. Maxwell (1999: 2-3) claims that 'native' peoples 'existed outside of the common bonds of humanity and the flow of history', making it possible for Europeans to exhibit Africans and other non-Europeans as museum displays, both drawing upon and reproducing the image of the African as not-quite human. This dehumanization was possible because the British looked at Africans not as Africans looked at themselves, but rather as a reversal of their own 'Europeanness', and ultimately everything that was not European (Achebe 1978: 2; Hammond and Jablow 1977: 16; Pieterse 1992: 30-39). Africans became viewed as child-like, in contradistinction to the attributes believed to make for a proper Victorian British (male) adult – having 'self-control, virtuous character, and rational mind', while Africans, like children were 'ignorant, impulsive, irresponsible, and without powers of reason' (Hammond and Jablow 1977: 64-65; Kuklick 1991). Europeans feared the claim Africans could make of an ancient kinship between Europeans and Africans, pushing Europeans to continuously distance themselves from the occupiers of the 'Dark Continent', while simultaneously frightening and thrilling them with the possibility of commonality (Achebe 1978: 4-5). In relation to this Eurocentric regard for Africa, Landau (2002: 4) writes of a constantly recurring, 'dizzying, self-devouring, "us–them" reversal'. Europeans saw their past mixed with Africa's present – they were our predecessors, our history, our selves before civilization.

Africans have long been represented as indicative of our past. Whether these are romantic notions of Africans and their supposed connections to nature (Pieterse 1992), or visions of an unchanging African, lacking a history and unable to cope with modernity, the impact is negative. As early as the sixteenth century, the idea was developed that the African was the 'natural man', living wild in his 'untamed nature'; by the nineteenth century the 'savage' African was a well established topos in European culture. And in the changing attitudes of the nineteenth century, not only were Africans regarded as living close to nature, but they were increasingly likened to animals (Pieterse 1992: 34f).

African cultures have long been understood as 'traditional', and while the idea of tradition is not necessarily presented in a negative light, it is at best a double-edged sword. At its worst, it has been used to provide the foundation, for example, for South African apartheid:

> Images are conveyed of a 'traditional' African culture that is unchanging, homogenous and communal – just as it always was in some mythical past – as compared with a 'modern' white culture that follows an historical trajectory of dynamism, diversity and individualism. Used in this way, 'traditional' culture and 'traditional' economics are not simply descriptive labels. They are part of an argument calling upon and reinforcing deep-seated images of an imagined past in order to account for marked discrepancies in wealth between whites and Africans in terms of supposed obstacles that 'traditional' beliefs and practices put in the way of progress and development (Spiegel 1994: 187).

Not only are these images used to explain poverty, but also to justify it. People who are stuck in the past certainly don't need the same complicated amenities as we do – simple, historically-based and basic alternatives are seen as adequate for people who represent our own past. However, even worse is that accompanying these justifications for keeping the poor poor is the sense that the poor and the 'traditional' know how to suffer, not feeling physical or emotional pain as deeply as the rich or Westerners do. As Albert Schweitzer, famed doctor to Africa, once said, 'The African is indeed my brother but my junior brother', and thus deserved only junior medical treatments and amenities. According to Achebe (1978: 8-9), this 'white racism against Africa is such a normal way of thinking that its manifestations go completely undetected'. Because the West does not see this racism, Westerners cannot see, and consequently do not address our role in impoverishing 'Others' and our need to fix large inequities and problems. Once again the view of the 'Other' as not-quite 'us' validates further injustices against 'them.'

In sum, nineteenth-century Euro–American culture constructed an idea of Africa and its peoples as primitive, savage and child-like, traditional and unchanging, living outside of the process – and progress – of history, close to nature. It is an idea so pervasive, so seductive in the way it rationalizes the Western hegemony, that it persists virtually unchallenged to this day. Any efforts that contradict these myths are filtered out or marginalized: thus the urban African is suppressed in favour of the rural; the modern in favour of the traditional; and so on. What we call the 'real' Africa is actually the Africa that exists in our collective imagination: rural, traditional, primitive, exotic, colorful, helpless and totally vulnerable to outside forces. As we shall discuss, the idea of Africa that served colonialism so well endures in popular culture, not least in some of the best known and celebrated photographic images in the West.

Photography in the Service of Colonialism

If the myths of Africa were already well formed by the middle of the nineteenth century, then the advent of photography served to confirm the expectations of European audiences.

Almost from its outset, photography was employed toward the goal of social and racial classification; it became a tool with which the hierarchies of race and class were recorded, analysed, and so confirmed, both in Europe and the colonies. Already in nineteenth-century Europe, prisoners were photographed, not only for the purposes of individual identification, but also for the general evidence the pictures revealed about the criminal face. The 'Other' became known through the visual record, and in so doing provided a template for defining the self (Hamilton and Hargreaves 2001; Pieterse 1992: 30-33; Tagg 1988).

As early as the 1860s photography was used in the process of anthropometry, the physiognomic method of measuring the human body, in which the colonial subject was photographed naked against a measuring stick or grid (Pieterse 1992: 46-47). Social Darwinists carefully measured and recorded the characteristics of the 'primitive' races, which they incorporated into a hierarchy of class and race, in anticipation of what was regarded as the inevitable and the inexorable decline following contact with the 'successful, energetic' Northern Europeans (Hamilton and Hargreaves 2001: 75-76; Markus 1990: 37; Maxwell 1999: 40-51; Pieterse 1992: 51).

The idea that some races were doomed to disappear, while others were fated to prosper and multiply, was a continuation of the discourse that had been devised in the seventeenth and eighteenth centuries to legitimize the large-scale genocide associated with European settlement of the New World. The crucial difference was that by the 1860s, extinctions were no longer attributed to divine providence, but were explained as the inevitable outcome of the natural law of survival of the fittest. The 'doomed' races were identified as relics of an earlier stage of humanity. They were considered valuable as shedding light on the path as the fitter races progressed to civilization, but they were also judged too fragile to survive the effects of competition. It was calculated that their rate of decline was greater than the time available to overcome the temporal distance between savagery and civilization (Maxwell 1999: 49).

Despite the claims of colonial photographers as to the 'infallible truthfulness' of the medium, then, images of Africa and the colonies were not neutral. They were determined by the experiences and tastes of their audiences, and by the assumptions they shared with the photographers who supplied the images. Colonial images decontextualized their subjects, making them 'available' to their Victorian viewers for the staging of 'eternal truths' (Osborne 2000: 17-21), disguising 'cultural assumptions and political visions' as natural law (Ryan 1995: 62). Colonial photographs were interpreted so as to confirm 'the European's sense of superiority', alternately filtering out or emphasizing the disturbing and exotic qualities of the 'Other' (Osborne 2000: 46).

Africa in Contemporary Imagination – Afro-Romanticism

The colonial view of Africa operates through a series of binary oppositions – our culture to their nature, modern to primitive, civilized to wild and so on – that persist in contemporary imagery (Bruner and Kirshenblatt-Gimblett 1994), and within which, in a broader sense, the NGO can be said to operate. Indeed, it is striking just how deeply the myth of Africa informs contemporary culture. It resides in movies, on TV, in the news media, airline flight magazines, children's books, in foreign policy and academia, from the board room to the school playground.

A recent article about a music festival in Mali in the *Telegraph Magazine*, the weekend glossy supplement to the broadsheet *Daily Telegraph*, begins as follows:

> We are two hours out of Timbuktu, jolting along in battered Land Cruisers, when we spot our first Tuareg – about a dozen of them, 100 yards ahead, sitting tall on their swaying camels. *We have seen plenty of town-Tuareg, turbaned jewelry salesmen hanging round our hotel. But these are the real thing*: veiled men of the desert, swords and rifles hanging at their waists. And glimpsed against the thin dune scrub, *they have an appearance of unnerving potency* – like Red Indians on the horizon, Hell's angels in the rear-view mirror or even Tolkein's black riders.
>
> *The Tuareg, the 'blue men', veiled lords of the Sahara, are one of the world's great romantic peoples* (Hudson 2003: 38, emphasis added).

Here is the image of a people unchanged, dangerous, raw and untamed, potent, close to nature and – unlike the urban, turbaned jewelry salesmen – 'the real thing'.

The myths of Africa stifle even the best efforts to create alternate frameworks within which to understand Africa and our relationship to it. Indeed the romanticization of Africa is a phenomenon that is celebrated at the highest levels. As recently as 2000, the United Nations honored the work of Angela Fisher and Carol Beckwith, authors of several best-selling, coffee-table texts, including the famous *African Ark*. Yet, like the *Telegraph* article, the imagery of Fisher and Beckwith is pure Afro-romanticism, perpetuating the most abiding myths of the continent and its peoples. Most commonly promoted as 'preserving timeless rituals' of 'cultures in crisis', Fisher and Beckwith have spent several decades traveling across the African continent photographing 'traditional' ceremonies. Their work has been published extensively in books, posters and postcards, and *National Geographic*. It provides arguably one of the foremost public faces of Africa in the West.

Fisher's and Beckwith's often stunning images portray an Africa rich in cultural diversity, colorful, energetic and vibrant. Yet beneath the surface, the meanings they convey are little changed from colonial times. Within the 'accumulated reservoir of sense', the imagery fulfills the expectations of the audience, who seek to be delivered to distant and exotic times and places, especially those peddled as verging on extinction.

Robert Morton's introduction to Fisher's and Beckwith's retrospective book, *Passages*, published in 2000, is worth quoting at length, for embedded within it, as within the pictures themselves, are many of the deepest and most enduring myths about Africa. Morton writes that the so-called 'journeys of the spirit', that is the rituals and ceremonies that Fisher and Beckwith photograph:

> ... enable people to rise above the mundane happenings of everyday life and connect them with larger forces, higher powers. *Nowhere in the world today has this been more true than in Africa.* All over the continent ... boys and girls, men and women, gather together periodically to participate in ritual acts that enhance their lives, *connecting them with one another, with their ancestral traditions, and with nature* ... Beckwith and Fisher know very well that much of Africa has come fully into the twenty-first century. But they have left the modern world to photojournalists. Their aim has been to document as fully and artistically as possible the *traditional rituals that persist more or less unchanged, even in the modern world.* It is fortunate indeed that they have done so, for *many of these rites are dying out or becoming altered* as Africa assimilates the habits and products of other parts of the world ... Perhaps their greatest gift to the future, however, will be '*the remarkable artistic document of their photographs*, preserving the rich and varied ways of life that have sustained *millions of people who live close to nature in communities whose shared values and common spirit could serve as models for all the world*' (Fisher and Beckwith 2000: 3, emphasis added).

These comments are so singularly lacking in reflection that it feels, on the evidence of this passage at least, that not much has changed in the last century and a half, either in the world of the viewer or that of the viewed.

As does the *Telegraph* 'Tuareg' excerpt, Morton twice reprises the myth of the African close to nature. This is the Africa most appealing to the West, the tourists' Africa. As Spiegel (1994: 191) notes, there is not necessarily any political agenda associated with the recycling of this idea. There is, however, a 'distinct lack of self-conscious evaluation of the political implications of the images conveyed'.

Morton's text separates the urban from the rural, and equates them with the modern and the 'traditional', respectively. The 'real' Africa is not modern. Rituals and cultures are primitive, static, and unchanging; and the essential Africa – the Africa avoided by photojournalists – is untouched by the outside world. It is the world view within which the 'authentic' Tuareg wears blue robes and rides a camel across the desert; in which a 'real' Maasai resides in the village, wearing traditional red robes and beads. We cannot envisage an urban Maasai – or if we do, she would seem out of place. (On this, see for example Bruner and Kirshenblatt-Gimblett 1994) There is the traditional, rural African, and the modern, urban African, and only the traditional African seems to have value.

The value of Fisher's and Beckwith's images is imputed to be their ability to preserve disappearing cultures and practices (the same exigency that was registered in the need to photograph native Australians facing 'extinction' at the turn of the twentieth century). In suggesting that it is contact with the 'modern' West that precipitates change, therefore, Morton not only denies the possibility of change from within, but also, one presumes unwittingly, recalls nineteenth-century Social Darwinism: once contact has taken place, the 'traditional' (read 'primitive') culture will inevitably disappear – and must therefore be preserved by the camera. The darker outcome of this is that as the photograph becomes the reality (per Sontag), so the 'original' becomes dispensable; the culture has been 'saved' for posterity by the photograph.

Thus for Morton the camera functions through the cultural conventions of photography as truth, and denies any cultural perspective of the photographer or the viewer. His view is based on the idea of the camera as objective witness, of the 'infallible truthfulness' of the photograph. In fact, the pictures themselves both draw from and build upon the reservoir of sense for their meaning. The images say more about our taste, about our attitudes towards Africa, and ultimately about how we view ourselves, than they do about the people and ceremonies they portray (Osborne 2000: 46; Pieterse 1992).

Famine, Disaster and War – Afro-Pessimism

What applies to glossy coffee table books can also be said for those texts intended to raise awareness of the terrible events most often associated with the African continent. Thus another component within the 'reservoir of sense' that was used to justify both the trade in human beings and imperial expansion was the idea that Africa is the site of our primitive ancestry, now inhabited by a barbaric and helpless population. While the place name might change – from Biafra to Rwanda – the underlying premise does not. The names are contiguous and conflated with the idea of disaster. Africa is a place of drought, famine, and war.

Some of the most powerful images of famine are those of Sebastião Salgado (1986), taken in the Sahel during the drought of the mid-1980s. Salgado's photographs of famine and starvation are highly aesthetic images, appearing in prestigious art spaces and books, and as major photo-essays in the 'serious' news magazines.

The ability of the camera to beautify the downtrodden has a long tradition (Sontag 1979: 110-12). Recently, however, the type of imagery exemplified by Salgado's *Sahel. L'Homme en Detresse*, a book of famine images he contributed to Médecins sans Frontières (MSF), has come in for criticism (Benthall 1993). In defense of what has become known as famine/disaster pornography, Salgado disavows himself of the conjunction between famine imagery and aesthetics:

> I never go to do a good picture. What is a good picture? No. I go to stay inside
> my story, to try to understand what's going on, to be close to the people I
> photograph, and to create a flow of information that we can use to communicate
> something (quoted in Light 2000).

Yet Salgado is clearly interested in composition – in making 'good pictures'. In
keeping with the ideals of 'high art', his pictures are intended to stand alone, un-
mediated by any explanatory texts, their messages the apparently self-evident
testimony of the camera lens. Salgado's photographs are so carefully composed,
beautifully lit, and so replete with visual references, that to deny their aesthetic
component is to deny the very root of their power. Salgado's 'stories', replete with
religious references, address the great themes of Western art – pain and sacrifice,
heroism and villainy. They are iconic images populated with iconic figures.

In *L'Homme en Detresse*, the Westerners, who are all white people, are
presented as icons of salvation. They evoke the heroes of nineteenth-century
history paintings and the Christian narrative. MSF doctors are identified by name,
and usually by place of origin too, in the captions; they are individuals, with whom
we are encouraged to identify. The Africans, on the other hand, are the nameless,
and often faceless figures who serve as the objects of Western beneficence. Thus
the characters in the 'story' are reduced to types – the starving, dependent African
and the heroic European.

At a minimum it is questionable how 'close' Salgado can get to the people
he photographs, given power differentials, time limits, language and other
cultural barriers, and so on. However, more insidious is how the famine story
– like the work of Fisher and Beckwith – draws upon a long history of western
imperial notions. Famine pornography is morally repugnant, yet even beyond the
extraordinary aesthetic qualities of his pictures, Salgado's 'story' is not original. It
is that of the down-trodden African, incapable of self-improvement, and reliant on
Western intervention to save his or her self from a self-inflicted failure to progress
or thrive (On this, see for example Benthall 1993; Moeller 1999).

The question is whether the good these pictures achieve is greater than, or
outweighed by, the power the images have, within the reservoir of sense, to reinforce
the pre-existing Western imagination of what Africa is. Salgado acknowledges
that he has a responsibility to his subject matter:

> I hope people look at my pictures because my pictures tell the story accurately.
> I try to link my pictures with an historical moment that we are living. I believe
> that these pictures will stay or they will disappear depending on whether or not
> they ultimately are linked with the historical moment. It's not because they are
> good pictures or bad pictures (Light 2000).

But it is because they are good pictures; it is the aesthetics that carry the images.
The meanings of images are not fixed; they cannot remain linked only to the
historical moment. Rather, how we understand photographs is informed by a range

of factors including a historical reservoir of sense, and the context and historicity of the moment in which they are viewed. Sontag (1979: 105-6) speaks eloquently to the fragmentary qualities of the photographic image. The meaning of the image changes according to the context in which it is presented, and so since the context is continually changing, the meaning is never secure. Meanings become eroded, 'bound to drain away'; truths become relative. 'One of the central characteristics of photography is that process by which original uses are modified, eventually supplanted by subsequent uses'. Images today are so rapidly commodified into popular culture that their moral half-life has become perilously short. Moreover, they are rapidly undermined by their artistic qualities. Mitchell (1994: 309) puts it succinctly: aestheticization is 'a continuing cover-up of evil under the sign of beauty and rarity' (see also Sontag 1979).

Working to Change the 'Reservoir of Sense'

The NGO is not immune from the 'reservoir of sense'. NGOs inhabit the same ideological spaces as Beckwith and Fisher, as Salgado, as newspapers, TV, movies, advertising, and popular culture in general. And, until quite recently, even those organizations dedicated to the 'development' of Africa have functioned at best unconsciously so as to perpetuate the same basic conceptions of Africa. As Jonathan Benthall notes, the 1981 Save the Children poster of an emaciated black hand supported by a well-fed white hand makes the same assumptions as we have suggested here for *L'Homme en Detresse*, in promoting the idea of a Third World dependent on Western charity. Benthall describes the Save the Children poster as racist (Benthall 1993: 179).

In the way that NGOs bring these images to potential donors and the public in general, they are at best unwittingly participating in the perpetuation of negative stereotypes, tapping and replenishing the 'reservoir of sense'. Through the 1980s there was a growing sense of responsibility among the NGOs to the ways in which they projected their mission, resulting, in 1989, in the adoption by the General Assembly of European NGOs of a 'Code of Conduct on Images and Messages Relating to the Third World'. In essence, the guidelines focus on the avoidance of demeaning imagery and stereotypes, on respecting the rights of those depicted, on giving voice to those traditionally rendered silent, and on highlighting the causes of poverty. In 1991 the Save the Children Fund UK instituted similar guidelines for all personnel involved in image production (Benthall 1993: 181-184).

During the 1990s, UNICEF developed an 'umbrella' approach to image production designed to protect 'identity in emergencies and when covering controversial and culturally-sensitive issues such as child labour, sexual exploitation, gender violence or discrimination, recruitment of child soldiers, trauma and the impact of HIV/AIDS' (Nsubuga 2003 pers. comm.). In the case of the author's experience with UNICEF Zambia, these guidelines played out in the field through the relationship between field worker and photographer who jointly ensure that the safeguards remain in place.

As many people acknowledge, ourselves included, one unintended consequence of this process is the production of some very bland imagery (Benthall 1993: 185-186). In conforming to the latest standards, imagery can become not only culturally sensitive, but also artistically muted, and hence less potent as fundraisers. More fundamental, however, is the question of whether what we might call politically correct imagery can have any lasting effect on the vast reservoir of cultural knowledge. Our ideas of Africa are so deeply embedded that it could take generations of re-evaluation to overcome.

One solution, particularly popular in the last decade or so, to the question of how to represent Africans fairly, has been to give the camera away – that is, to 'allow' Africans to photograph themselves. Perhaps most laudable among these efforts is the 'Rwanda Project', in which orphaned children used disposable cameras to document their daily lives, often to great effect. It appears, on the surface at least, to give voice to those traditionally not heard. Yet even this remarkable endeavour fails to challenge conventional wisdom. In reality, the context within which the pictures are created and presented – indeed their very premise – remains no different from that of the professional photographer on assignment for the NGO – these are pictures of Rwandan orphans taken in order to raise funds for development projects. Moreover, the strength of the pictures is predicated on their authorship – that is, the photographic naïveté of the children, indeed their general lack of worldliness, suggests a visual purity, an essential 'truthfulness'. In other words, it is the great myth of photography. Yet even if one could support the claim that the images are somehow in closer proximity to the 'truth', as they pass through the various editorial processes that filter, aestheticise and re-contextualize, so the position of the images, within the prevailing reservoir of sense, becomes indistinguishable from those taken by the professionals. Indeed the fact that they are taken by local children, by masking even further the mediated qualities of the image, can only enhance their ideological value.

In a general sense, any pictures that are used to highlight a need will almost by definition reinforce the stereotype of the needy. So we must try to deliver a more complex picture of Africa. Both words and images must be anchored in a more honest context – historical, political and economic. Text and pictures must be put together in a more sophisticated way so as to create more complex, reflective documents.

One recent publication does, we suggest, offer some hope. Fazal Sheikh's *A Camel for the Son* (2001) profiles Somali refugees, mostly women and children, in Kenyan relief centers. The author spent over a year interviewing and photographing. The pictures are interwoven with a text that gives voice to the individuals depicted; we hear the people whose faces we see in the book. The book distances itself from many of the conventional tools of the documentary photo-essay that privilege the visual, emphasize formal qualities and incorporate iconic imagery. In contrast, *A Camel for the Son* uses text and image together to weave cultural, political and historical contexts into the story of the Somali refugees. The audience is given the opportunity to understand the reasons behind the disaster, and, through the interviews, learns that the misfortunes of the refugees were not part of the natural cycle of things. The peoples' refugee status is not a natural function of their 'role' as generically poor people of the Third World.

A Camel for the Son further departs visually from conventional disaster narratives by portraying physically healthy individuals. Without denying the ravages of the Somali conflict and the deadly journeys across the desert, the survivors are photographed after they have recovered physically from the ordeal. Sheik's photographs do retain certain artistic conceits – such as the use of black and white, the staged formality of the compositions – that compromise the otherwise progressive qualities of the work. Yet overall the photographs are less likely to reinforce conventional disaster images by which we have come to know these issues. The heroes of this story are the victims, not the NGOs. They are portrayed as people of intelligence; they are given agency.

As Sheik's work appears to suggest, if we are to bring about meaningful changes in the relationship between Africa and the West, to create one that is more equitable, less exploitative, a partnership of equals, then we need to create an image of Africans as capable, cultured, and intelligent – as well as frequently compromised by those structural disadvantages exploited by their own governments, foreign governments, and other political, commercial, economic and cultural institutions. In part, this requires a more frank admission of how we have benefited, materially and culturally, from our relationship with Africa, and of how our constructed idea of Africa helps maintain the status quo. More sensitively described images of the suffering African do little in this regard. We must find 'Africanisms' other than the traditional, the exotic, the romantic, the primitive and natural; we should cover stories other than famine, war, and drought. As Moeller (1999) observes, we need to cover the stories that lead to disaster, before the disaster happens. We must continue to question how the reservoir of sense drowns out the ability for those poor and disadvantaged people who really are out there to become empowered. And perhaps most importantly, we must create stories that implicate us in the creation of a dependent Africa, exposing our dependency as much as theirs.

References

Achebe, C. (1978), 'An image of Africa', *Research in African Literatures*, Vol. 9, No 1, pp. 1-15.
Barrell, J. (2000), 'Death on the Nile: Fantasy and the Literature of Tourism, 1840-60', in C. Hall (ed.), *Cultures of Empire, A Reader: Colonizers in Britain and the Empire in the Nineteenth and Twentieth Centuries*, Routledge, New York, pp. 187-206.
Barthes, R. (1977), *Image, Music, Text*, Hill and Wang, New York.

Benthall, J. (1993), *Disasters, Relief and the Media*, I.B. Tauris, London and New York.

de Botton, A. (2002), *The Art of Travel*, Pantheon, New York.

Bruner, E.M. and Barbara Kirshenblatt-Gimblett, (1994), 'Maasai on the Lawn – Tourist Realism in East-Africa', *Cultural Anthropology*, Vol. 9, No. 4, pp. 435-470.

Fisher, A. and Beckwith, C. (2000), *Passages: Photographs in Africa*, Harry N. Abrams, New York.

Hamilton, P. and Hargreaves, R. (2001), *The Beautiful and the Damned: The Creation of Identity in Nineteenth Century Photography*, Lund Humphries (in association with the National Portrait Gallery), Aldershot.

Hammond, D. and Jablow, A. (1977), *The Myth of Africa*, The Library of Social Science, New York.

Harvey, D. (1996), *Justice, Nature and the Geography of Difference*, Blackwell, Oxford.

Hudson, M. (2003), 'Desert Blues', *Telegraph Magazine*, 28 June 2003, pp. 38-45.

Kuklick, H. (1991), *The Savage Within: The Social History of British Anthropology, 1885-1945*, Cambridge University Press, Cambridge.

Landau, P.S. (2002), 'Empires of the Visual: Photography and Colonial Administration in Africa', in P.S. Landau and D.D. Kaspin (eds), *Images and Empires: Visuality in Colonial and Postcolonial Africa*, University of California Press, Berkeley, CA, pp. 141-171.

Light, K. (2000), *Witness in Our Time: Working Lives of Documentary Photographers*, Smithsonian Institution Press, London and Washington.

Markus, A. (1990), *Governing Savages*, Allen & Unwin, Boston.

Maxwell, A. (1999), *Colonial Photography and Exhibitions*, Leicester University Press, London.

Mitchell, W.J.T. (1994), *Picture Theory: Essays on Verbal and Visual Representation*, University of Chicago Press, Chicago.

Moeller, S.D. (1999), *Compassion Fatigue: How the Media Sell Disease, Famine, War and Death*, Routledge, New York and London.

Mudimbe, V.Y. (1994), *The Idea of Africa*, Indiana University Press, Bloomington and Indianapolis.

Osborne, P.D. (2000), *Travelling Light: Photography, Travel and Visual Culture*, Manchester University Press, Manchester.

Pieterse, J.N. (1992), *White on Black: Images of Africa and Blacks in Western Popular Culture*, Yale University Press, New Haven.

Ryan, J.R. (1995), 'Imperial Landscapes: Photography, Geography and British Overseas Exploration, 1858-1872' in M. Bell, R. Butlin and M. Hefernan (eds), *Geography and Imperialism 1820-1940*, Manchester University Press, Manchester, pp. 53-79.

Salgado, S. (1986), *Sahel: L'Homme en Detresse*, Prisma-Presse.

Sheikh, F. (2001), *A Camel for the Son*, www.fazal sheikh.org, accessed 22 April 2009.

Sontag, S. (1979), *On Photography*, Penguin, London.

Spiegel, A.D. (1994), 'Struggling with Tradition in South Africa: the Multivocality of Images of the Past', in G.C. Bond and A. Gilliam (eds), *Social Construction of the Past: Representation as Power*, Routledge, London, pp. 185-202.

Tagg, J. (1988), *The Burden of Representation: Essays on Photographies and Histories*, University of Massachusetts Press, Amherst.

Thompson, J.B. (1984), *Studies in the Theory of Ideology*, University of California Press, Berkeley, CA.

Williams, R. (1973), *The Country and the City*, Oxford University Press, New York.

Wolf, E.R. (1999), *Envisioning Power: Ideologies of Dominance and Crisis*, University of California Press, Berkeley, CA.

Chapter 5

The Bulimic Consumption
of Pygmies: Regurgitating an Image
# of Otherness

Stan Frankland

The Allure of the Other

Tourism relies on difference, on the provision of the extraordinary. We all want something more out of our existence than the mundane familiarity of our daily working lives. As a means of escaping the rigours of regularity, the holiday has become a concentrated life moment when we can actively engage in pursuits beyond whatever constitutes our normal practice, even if this means doing nothing at all. The experience of tourism is that all too brief moment in our regulated patterns of time when we can revel in the atmosphere of something beyond the bounds of the everyday (cf. Urry 1990), when we can enjoy something Other. This sense of Otherness can be anything, either material or immaterial. It can be something as banal as the sun; as elusive as the uncertain desire of a sexual encounter; as trifling as a tourist trinket; as exotic as an alien environment; or indeed as arcane as an encounter with some strange being. Whichever or whatever combination of Otherness we may choose to partake of, the aura of authentic difference can always be found shimmering in the background. Even if we wrap ourselves up in the tourist bubble as a protection against the very fear of difference (cf. Boorstin 1964), we still consume that Otherness, albeit in easily packaged and manageable forms. So long as we can believe in the authenticity of the experiential encounter, so long as we feel that we are getting something above and beyond the normalcy of home life, then it can be said that we have encountered difference.

Clearly, there are multiple ways in which such experiences can occur. The overarching idea of Otherness has been translated to fit across the broad spectrum of contemporary tourism styles. From package holidays to ecotourism, all fashions of travel incorporate at least some aspect of this vague yet crucial motive force, the differentia being the degree of difference desired itself. The seductive quality of the dissimilar reaches its apotheosis in what has become known commonly as 'ethnic tourism', the form of travel in which we engage directly with those who may differ physically or whose social practices are significantly alternative to our own. In this context, the idea of difference is paramount, it being the basic Otherness of those people being visited that makes them a tourist attraction in the first place. Such a style of tourism may seem strangely at odds with the sophisticated industry of today, nevertheless it continues to be a key component of the business, actively marketed and avidly consumed, particularly within 'underdeveloped' countries and among 'indigenous' peoples.

Van den Bergh (1994: 10) has described this 'quest for the Other' as 'the last wave of capitalist expansion into the remotest periphery of the world system', a wave that crashes into the people and places left beyond the wash of modernity. By being sucked within the boundaries of the pleasure peripheries of tourism, environments and ethnicities have been ascribed a use value within the patterns of the consumption of difference, a commodification that rests on the maintenance of Otherness. Through an ideology that equates change with loss, the consequences of the expansion of touristic space on the inhabitants of these areas have been perceived as being almost entirely negative. This proselytising position can be summarized by a simple formula: tourism causes dependency; dependency de-authenticates culture; culture is destroyed by tourism. Here lies the teleological conceit at the heart of travel motivated by the desire for radical Otherness. Any Eden found must soon be lost, precisely because it has been discovered; any Adam or any Eve will soon be polluted, their purity compromised by the very act of contact. This paradox is particularly acute when the ethnically demarcated Other is transformed into a spectacle. Within this 'paradox of the primitive' there is both a recognition and fear of the destruction wrought by the Self on the Other. In this sense, MacCannell's (1976) concept of 'staged authenticity', ranging from the wholly fake front stage to the fabled back stage of ultimate authenticity can be recast as a sliding scale of contamination by which the relative merits of the object to be consumed can be judged. The closer to the front, the greater the assumption of pollution and commercialisation; the nearer to the back, the greater the perception of authenticity and purity. There is a mammonistic perception of wealth as an evil and corrupt influence on the Other combined with a cognisance that it is the tourists themselves who are culpable of instilling the contaminating desires for modernity.

Although this argument has become a familiar *cri de coeur* within the literature on tourism, the self-righteousness and self-loathing that lies behind this attitude is not confined to the academic critics of tourism alone. Rather ironically, it is commonplace among those people who sell and consume ethnic tourism. Laments about the commercialization of the 'ethnic' object are found frequently within guidebooks such as *The Lonely Planet* and these are repeated by the actual tourists themselves, normally after the disappointment of encountering an Other who fails to meet their expectations (cf. Frankland 1999). Yet despite these widespread reservations about and critiques of ethnic tourism, it remains a popular and ongoing component of the global tourism industry. Indeed, as tourism has expanded to incorporate ever more people and places within its fuzzy boundaries, the very idea of Otherness has become an increasingly beguiling commodity. The imagined scarcity of environments and populations untouched by the grasp of globalization continues to be an appealing notion, a fascination which, in turn, reinforces the paradox of the primitive.

It has been said that for tourists the myth of the 'authentic savage' acts as a primary luring device, serving as a sign for the pure and original Other untouched by the horrors of the modern world (Bruner 1991). Regardless of the reasoning behind this touristic desire, there are numerous groups of people dotted across the planet who are integrated within the parameters of tourism, willingly or not, precisely because they are thought to correspond to this fantasy. There is, in effect, a catalogue of commodified difference from which to choose a suitable version of the 'primitive' to visit, a list that includes such people as the Australian Aborigines, the Inuit and the Amazonian Indians. One of the most potent brand names included on this inventory of idealized existence are the Pygmies of the central African rainforest. The desire to find some kind of authentic Pygmy repeats the sentiments that lie at the heart of ethnic tourism and the emotional search for a lost Self in a pure and uncontaminated Other. Indeed, the very name Pygmy acts as a symbolic device, triggering a collage of eidetic images, vividly clear recollections drawn not from reality, but from the imagination instead. Such indexing and dragging (Rojek 1997) creates a montage of the exoticized Other comprised of generalized fantasies of the 'primitive' and specific romanticizations of Pygmies as living embodiments of life as it was lived in the Stone Age (Frankland 2001). The combination of fetish and phantasm provides a 'primitivist cosmetic haze' (Carey 1992: 36) that sustains the notion of authentic difference and nourishes the touristic desire for Otherness. Through this eidotropic confusion, the Pygmies have become enduring objects of fascination within the teratology of tourism.

The Myth of the Pygmies

In order to understand how the Pygmies have been transformed into symbolic beings within the tourism industry, it is necessary to return to the moment of their 'discovery' during the colonial scramble for Africa. Images of the Other have run as a constant thread through the European past (Jahoda 1999: 1), and the Pygmies had long been mythic figures within the scientific and philosophical debates concerning what constituted humanity. As a classic illusion of the unknown Other, the Pygmies were one of the corpus of 'imaginary beings' believed to inhabit the world beyond the civilizations of the Ancient Romans and Greeks (Borges 1970: 188). However, unlike the other fabulous beings, the Pygmies became a proven reality when they were finally 'discovered' deep in the 'unexplored' heart of Africa by Schweinfurth, in 1870. As a consequence of this meeting, Schweinfurth ascribed to them the dormant mythic formation of the Pygmy derived from the Ancients: 'at last, was I able veritably to feast my eyes upon a living embodiment of the myths of some thousand years!' (Schweinfurth 1873: 127). Memory and fantasy combined to recreate living beings as an image of antiquity. It was this moment, when the remote in time were fused with the remote in space and given a corporeal form in the here and now, that marks the 'invention' of Pygmies (Bahuchet 1993). From that time onwards, the generic naming has stuck with them repeated *ad infinitum* within the Western consciousness of Africa. Similarly, the concatenation of myth with the modern was retranslated into an accepted 'narrative history' (Clifford 1997: 319) of factual verity through the writings of the many travellers, explorers, hunters, missionaries and colonial officers who recorded their experiences in Africa. The Pygmies became an authorized reality and in so being were transformed into 'utter savages of scientific value' (Rosaldo 1982: 309-325) and also into perpetual primitives within popular culture. It was the beginning of a very contemporary mythology.

On one level, this can be thought of as another example of the assimilation of difference through the act of categorization. Tomas (1996: 99) takes the naming, marking and collecting of difference to signify the transition from the marvellous to the scientific. However, the 'discovery' and classification of the Pygmies did not make them any less of a mythic entity. The more arcane elements of the Plinian teratology may have been discarded, but the idea of extreme Otherness remained in place within a refined morphology of primitivism that revolved around two conflicting conceptions of evolution. The first understanding saw man progressing, albeit fitfully at times, from a state of savagery towards the zenith of civilization along a unilineal path of evolution. The second perspective held that man had always been much the same and that the so-called savages were nothing more than the 'degenerate descendants of far superior ancestors' (Lubbock 1875: 465). The fundamental question surrounding the updated myth of the Pygmies remained the same: just what does it mean to be human? However, the explanatory framework supporting the answer differed, incorporating the diverse ideologies born out of social Darwinism. Maffesoli (1996: 7) has written, 'each era hauntingly repeats multiple variations on a few familiar themes' in which the only thing that differs is the angle of approach. As such, while there is change within the make up of the myth so too is there a basic continuity, that being the core thematic of Otherness.

Here, there is a correspondence to Shields's (1991: 61) notion of place-myths in which certain associated images lose their significance through time and become 'dead metaphors'. While there is an addition of elements to a myth, there is also a correspondent decay and redundancy of other components. Through this process of loss and accretion, a 'strange mutation' (Barta 1995: 242) takes place that alters both the content and the context of the particular mythic formation to fit a particular situation or way of thought. In this context, we can think of the Pygmies as having been brought within the boundaries of 'white mythologies' of the nineteenth century, a 'cultural explosion' in which 'texts from archaic and primitive cultures powerfully erupted into European civilization' (Lotman 1990: 379). The dynamic eruption of the very idea of Pygmies transformed them into a recognizable brand name within the larger discursive formation of the primitive. As a consequence of this dramatic event, the renewed myth of the Pygmies caught hold within the 'semiosphere' (ibid.). Within the affective milieu, the complex system of signs and the multiplicity of codes that generate thought and action, the symbolic figure of the Pygmy took on an effervescent quality that continues to enchant to this day.

Extraction and Return

The consequence of this 'cultural explosion' has been the lasting reproduction of the Pygmy as a logo signifying difference within the collective imagination. MacCannell (1992: 66) might think of the incorporation of the Pygmies into the global swarm of Euro-American myth as an example of 'symbolic cannibalism' that consumes and thereby eliminates difference. This stern view is redolent of the critiques of ethnic tourism referred to earlier and is, in part, reliant upon a notion of authenticity that has been lost. The original difference is imagined as having been a real and prior condition. However, as Root (1996: 68) points out, the 'new appreciation of cultural difference is all done with mirrors ... and, as in the past, what is usually available are the morphological forms that connote difference, which is to say, alluring commodities – difference in effigy, as it were'. As I have stressed above, the idea of Pygmies is and has always been based in the adaptive structure of a mythic formation. It is a metamorphic product of the imagination and cannot, as such, be eliminated.

It is possible, however, to retain the root metaphor of cannibalism in order to explain how the myth regenerates in form to provide an ever-shifting illusion of authentic alterity. Bauman (1993: 163-165 and 1997: 46-57) has theorized internal and external strategies for dealing with strangers by appropriating Lévi-Strauss's concept of the anthropophagic and anthropoemic suggested in *Tristes tropiques*. In Lévi-Strauss's (1973) original hypothesis, the anthropophagic strategy is the 'savages' method of nullifying the stranger's ontological Otherness via coercive assimilation, by devouring them symbolically. The anthropoemic strategy is 'our' way of dealing with the same problem. Instead of consuming the stranger, we exclude the danger by vomiting it out 'away from where the orderly life is conducted' (Bauman 1993: 163). Where Bauman differs from Lévi-Strauss is the separation of the inclusivist anthropophagic strategy and the exclusivist anthropoemic strategy between 'us' and 'them'. Rather, he suggests that both strategies co-exist as indispensable mechanisms of social spacing and that they are only effective when they offer a genuine 'either/or' choice. By reappropriating Bauman's reappropriation, it is possible to envisage a progressive circularity in which the two are combined into what can be called a bulimic consumption of strangeness that emphasizes the need to maintain a notion of extreme difference, not to destroy it. By this, I mean that strangeness or Otherness, or whatever else you may want to call the perception of difference critical to the definition of the Self, is still actively consumed, but after the feast, it is regurgitated. It is perhaps difficult to think of vomit as an appetizing dish, but the puked out remains of difference are then absorbed into the menu of choice for identity formation and the appetite is whetted once more for the outward desire to consume over again. In this sense, vomit has a regenerative quality. It does not mark the end of the process of consumption; instead it signifies the beginning of a new phase.

Despite the somewhat florid nature of this theoretical metaphor, it does provide an idea of the repetitive pattern in which the Pygmies have been produced and consumed. Since the moment of their invention and their accreditation with the aura of ancient history, the myth of the Pygmies has been imposed, collected, brought home and displayed. However, it is not only the myth that is caught up within this progressive circularity, it is the people themselves. In the 1870s, the decade of Schweinfurth's 'discovery', an increasing number of explorers penetrated the uncharted depths of the African rainforest belt. In 1873, an Italian called Miani purchased two young Pygmies with the intention of bringing them back to Europe (de Quatrefages 1895: 182-183). Miani died during the return trip, however, the two Pygmies, Tebo and Chairallah, did make it back to Italy where they were adopted by Count Miniscalchi Erizzo, the then President of the Italian Geographic Association. At first, the pair were educated, both learning to speak Italian and Tebo being taught how to play the piano (ibid: 183). However, at the same time, a popular desire to witness these unique physical specimens led to their being promenaded in front of a curious public and discussed extensively in the Italian media.

In many ways, the story of Tebo and Chairallah is as significant an event as Schweinfurth's moment of 'discovery' because it marks the beginning of a process of physical and symbolic extraction and display that continues up until this day. In 1904, a group of Pygmies were collected and brought to America by the explorer Samuel Verner where they were exhibited among the other ethnic curiosities at the St. Louis World's Fair. The intention behind the Fair was the scientific demonstration of the stages of human evolution with the 'darkest Blacks' set off against 'dominant whites' and members of the 'lowest known culture' placed in contrast to 'its highest culmination' (Bradford and Blume, 1992: 94-95). A year later, a Colonel Harrison brought another group of six Pygmies out of the rainforests of Africa. They appeared at the London Hippodrome before being toured around the theatres of Europe where about one million people saw them perform. As an adjunct to their role as performers, they were taken to the Houses of Parliament and Buckingham Palace; photographs were taken of them and postcards sold (Figures 5.1 and 5.2); and, as in America, anthropometrists and anthropologists examined them as living specimen of evolution.

The production of the myth of the Pygmies had begun and anthropology was complicit within this manufacturing process. The scientific and the social were fused in the creation of the Pygmy as an example of 'savage splendour' during the ethnological show business of the nineteenth and early twentieth centuries. More importantly, however, the scientific fascination was the impetus for the development of the Pygmies as a spectacle in the burgeoning tourist industry of Africa. The object had been collected, consumed, interpreted, and simplified down to key symbols and these were then returned to the assumed source. This marks an intensification of the process of bulimic consumption in which ever increasing

Figure 5.1 *Colonel Harrison's African Pygmies*, **Scott Series No. 747. Published by Scott Russell & Co., Birmingham, 1905-1907**

Figure 5.2 *The Pygmies*, **Photographer W & D Downey, London. Published by J. Beagles & Co. London, 1905-1907**

numbers of visitors to Africa carried with them the myth of the Pygmies among their baggage of expectation and desire. Here was 'an Africa that was both paradise and wilderness, a place of spectacular but savage beauty' (Adams and McShane 1996: xii); but it was also an Africa populated by weird and wonderful manifestations of humankind. For much the first half of the twentieth century, there was a tourist circuit that took in a range of attractions that included the Virunga Volcanoes, home of the recently discovered mountain gorillas, the recently domesticated Pygmies, and also the giant Tutsis of Rwanda. This was a journey through time, travelling back along the great chain of being to the origins of mankind. The growth of Pygmic tours enabled tourists to visit a living laboratory of evolution in the heart of darkest Africa in which even the missing link appeared in the guise of the Pygmies (Frankland 2001).

After the war and the arrival of the commercial jet plane, the concept of tourism as a mass enterprise became more defined. There was a perceptible shift within the historical pattern of colonial tourism with the emergence of a more formal tourism space centred on National Parks and wildlife. With this came a more controlled form of international safari tourism and a reduced interest in the diversity of alternate spectacles outside of the Parks. As the 'wilderness myth' of Africa as 'a glorious Eden for wildlife' (Adams and McShane 1996: xii) became the dominant ideology behind tourism practice, the number of characters within the human zoo frequented by the earlier tourists reduced. The catalogue of commoditized difference shrank to include only the most recognizable of brand names, such as the Maasai and the Pygmies. However, the seductive allure of these peoples remained as potent as before. By the 1960s, a veritable Pygmy industry had grown up in the forests of northeast Zaire, Rwanda and western Uganda with numerous participants willing to dance and pose for tourists in return for money. Specific places, such as Epulu and the Semuliki valley, became established centres for these Pygmic tours and formed key nodal points in a tourist infrastructure that absorbed the regions into the broader network of tourism in East Africa (Frankland 2001: 246).

As time has gone by, the number and frequency of interactions between tourist and Pygmy has generally increased, the only break in the trend being caused by the outbreak of various wars. Yet as one group becomes unavailable to the tourist gaze, another picks up the mantle of primitive spectacle to the extent that there are now some groups of Pygmies spread out across the rainforest belt of equatorial Africa who have a history of tourism spanning nearly a hundred years. This rate of encounter has accelerated during recent years with the fragmentation of the tourism industry into multiple niche markets encouraging styles of tourism that have renewed the demand for the authentic. With a growing unease 'about the emptiness and commodification of mainstream "white-bread" culture, there are attempts to look elsewhere for meaning and cultural and aesthetic integrity' (Root 1996: 73) with the touristic quest for Other providing one of the main hopes for fulfilment. Once again, the myth of the Pygmy has mutated to respond to the changing needs of a 'spiritual capital' (Edwards 1996: 201) that underpins the tourist engagement with difference. There has been a change in discursive function with the once popular theme of evolution fading into the background as the paradigm of salvage ethnography resurfaces as the dominant trope, supported by the theme of the loss of primitive purity, the motif of the disappearing world and the fear of vanishing wisdom. A whole new cycle of extraction and return is underway, with the bulimic consumption of Pygmies maintaining a principle role in the regurgitation of Otherness.

The Intertwining Media of Myth

Throughout the process of mass production, the myth of the Pygmy has been refined within an ever-massing media. As a part of the progressive circularity of extraction and return, there is a powerful and dynamic repetition of representation that has moulded, remoulded and reinforced the myth of the Pygmies. The identity of the Pygmy as a fetishistic figure embodying Otherness was created and is recreated by and through complex patterns of imaginal repetition that generates the stereotype, the archetypal image that is then reproduced in numerous variations within the semiosphere. This mythologic system is comprised of all the various conduits of transmission, all the mythemes of translation and all the methods of consumption. Amongst these component parts, there are three distinct but interlocking levels on which this replication of the mythic Pygmy takes place: the textual, the aural, and the visual.

The Textual Level

This covers a massive range of printed material from both the popular and academic markets. It is the most prolific of the three levels, regurgitating the myth in a seemingly endless flow of texts. Unsurprisingly, one of the major mediums of transmission is the tourism literature of travel guides, travelogues and brochures, each format maintaining the notion of an authentic Pygmy as an exemplar of Otherness. The fantasies of the rainforest environment and hunter-gathering lifestyle are perfect ideal types to map onto the *terra incognita*, the remote of the twenty first century. The mythic figure of the Pygmy also appears in magazines as diverse as the *National Geographic* and *High Times*; in comic books such as Tintin and Tarzan; as both fact and fiction in books aimed at children; and in a plethora of novels such as Michael Crichton's (1983) *Congo*. In this realm of representation, the myth is less cohesive and more liable to variation as the motive of the market is less focused. Nevertheless, the same overall trajectory of bulimic consumption can be clearly identified. Whatever the text, the Pygmy remains encapsulated by the overall properties of the myth.

However, it is anthropology that has played the key role in defining exactly what form the mythologic takes as a result of the publication of Colin Turnbull's *The Forest People* in 1961. Although the Pygmies have been dissected by physical and human sciences alike, it was the romanticized vision at the heart of this best selling and often reprinted book that configured the Pygmy as a lasting and potent symbol within the semiosphere. In this work, Turnbull evoked an enduring vision of an unchanging Pygmy lifestyle in harmony with the rhythms of the rainforest, a classic fantasy of the pure, uncontaminated Other (Frankland 1999). He crystallized the underlying dualism that lies at the very centre of notion of the primitive; the deictic politics of 'us' against 'them' in which the chaos of modernity reflects the Hobbesian war of all against all while belief in the authentic Other represents a faith in the idyllic existence of the Rousseauan noble savage. And it has been this version of authenticity that has become the dominant trope within the mythologic system, endlessly reproduced to the extent that it has created, in effect, a static yet overflowing eidolon of the primitive isolate. As Roland Barthes (1973: 163) wrote, 'a myth ripens because it spreads' and it has been the power of this seductive image of purity that has influenced all those who produce and consume the ethnic object of desire.

The Aural Level

The musical Pygmy backs up the textual Pygmy, enhancing the myth on the aural level. From as far back as 1906, when 'Colonel Harrison's Pygmies' toured the vaudeville stages of Europe, the musical skills of the Pygmies have been manufactured as a commodity. Among the many souvenirs sold during their visit, there was also a selection of five phonographs that could be purchased by the inquisitive consumer (Green 1999: 169). This was the beginning of a process of production by both amateur and professional ethnomusicologists that has transformed the polyphonic singing and the music of the pygmies into a seductive sonic metaphor for romanticized difference. In the 1930s, Armand Dennis made sound recordings of the Pygmies of the Ituri Forest; in 1954, field recordings of the Mbuti made by Turnbull and Francis Chapman were published by the Smithsonian Institute (Grinker 2000: 95); and in 1966, Simha Arom and Genevieve Taurell released an LP of music by the Ba-Benzele Pygmies. As with the textual level, each act of collection freshens the myth thereby stimulating the next phases of production and consumption.

Although Turnbull's recording influenced the free form yodelling of Leon Thomas in the early 1960s (Grinker 2000: 95), it was Arom's LP that was the most significant in the mythologization of Pygmies as icons of pop culture. As Feld (1996) has observed in his discussion of 'Pygmy pop', from the moment of its release, Arom's record has influenced successive generations of musicians across the breadth of musical genres. Perhaps the most famous of these is the CD by Deep Forest on which certain refrains from Arom's work were sampled without permission, setting them against a lightweight techno rhythm. Opening with a portentous monotone about the ancient wisdom of the Pygmies, this record is the perfect example of the sonic encapsulation of the Pygmies within the locked groove of myth. Nevertheless, it proved so commercial that tracks from it appeared on over thirty different advertisements across Europe (ibid) and cropped up on various film soundtracks. In many ways, the Pygmies could be called heroes of world music, having inspired numerous cross-cultural fusions from the ethno-folk musings of Baka Beyond to the highbrow intellectualisation of the avant-gardist György Ligeti. On a recent CD, *African Rhythms*, Ligeti combined the complex polyrhythms and irregular multi-part vocalising of the Pygmies with piano compositions of his own. In a recent interview, the pianist of the Ligeti pieces, Pierre-Laurent Aimard, summed up the repetition of thought that lingers behind every realization of the Pygmy myth:

> It was extraordinary to find so-called primitive music had the complexity we ourselves were striving for. And it was a valuable lesson for us – for whom music has been commodified – to see how refined it can be when its function is purely social … we produce music on stage for listeners; they sing for themselves, and for the spirits. The Pygmies don't have the capacity to criticise and compare, they have no notion of professionalism. But as we know to our cost, professionalism can kill the artistic spark (Church 2003).

There is within this quote a clear connection to the tourist trope of loss, and a correspondence to the 'fascination with other cultures [that] has been a way for a certain type of aesthete to imagine an outside to the exhaustion and disasters of European culture' (Root 1996: 20-21). High culture or low, it is the same old story: the paradox of the Pygmy can shape-shift to fit any context at any time. Recorded and replayed for nearly a hundred years, the voice of the Pygmies is muted, silenced by the persistence of a myth that listens only to the beat of its own drum.

The Visual Level

In the same way that the sonic reduction of the Pygmies into a condensed variant of exoticized Otherness renders them voiceless, their incorporation within an equally potent visual media has the end effect of making them invisible. There is a significant history of cinematic and televisual representations of the Pygmies that replicates the fantasy of idealized essence that lies at the heart of the myth, reinforcing yet further the fascination with the difference that underpins our engagement with the world. This begins with their role as evolutionary throwbacks to the Stone Age in the early documentaries of such filmmakers as Martin Johnson, Lewis Cotlow and Armand Dennis. As with the forms of human display mentioned above, issues of race and civilization predominate which emphasize the 'childlike' qualities of the Pygmies. The genealogy of the filmic Pygmy moves on through their Disneyfication in a variety of cartoons that include *Mickey Mouse* and *Tom and Jerry*, and continues with their casting as modern day freaks in the 1960s pseudo-documentaries of the mondo genre of kitsch and schlock horror. The pictorialization of the Pygmy reaches on into the modern day and their appearance as a celebrated image of primitive purity in numerous TV documentaries, a format lent added authenticity through a symbiosis with the 'realist science' of ethnographic film.

Once again, the spectre of Colin Turnbull haunts the reproduction of the myth. Not only did he capture them on film himself, but also, after the success of *The Forest People*, he publicized his vision of Pygmy lifestyle on numerous TV chat shows (Grinker 2000: 78). As with the textual and aural levels, his intervention within the mythic formation of the Pygmies determined the trajectory it would take from then on: 'Turnbull's Pygmies' were and remain a classic invention of all that is imagined as good in the Other and lacking or lost in the Self. It is this aesthetic idiom that has become the sine qua non for all future representations of the filmic Pygmy. Each successive update of the myth begins with the same belief in 'romantic preservationism' (Rony 1996: 102) that removes the Pygmies from time and history. This cinematographic version of the 'ethnographic present' (cf. Fabian 1983) encapsulates the Pygmies within a frozen image of Otherness, an illusion of permanence enhanced by the regurgitation of a set pattern of representational strategies and visual tropes. Particular events and activities, such as hunting and honey collecting, are repeated from film to film, reproducing an exoticism of the body and the quotidian, and a reification of a romanticized state of nature. Together, anthropology and entertainment have combined to create a fetishized Pygmy, a 'savage icon' within the visual media.

However, it is not only through the essentialized Pygmy of film that the myth is reproduced on the visual level. As with the textual and aural mediums of myth, there are numerous channels through which the ideological aesthetic of the Pygmy is produced and consumed. This involves the direct visual engagement with the physical spectacle across the range of tourism experiences, both 'here' and 'there', 'home' and 'abroad'. Emerging from out of the long history of ethnological show business, the Pygmy has become a staple product within the museums of the world. The methods of display within the museological regurgitation of the myth are, no doubt, more sophisticated than the crude evolutionism of the past. However, through such visual representations as dioramas, sculptures, and exhibitions of Pygmy art and artefacts, a refined image of primitivism buttresses the projection of fantasy beamed onto the silver screen and through the cathode ray. If there is any difference, it is in the perception of authenticity that can be attributed as a result of the immediacy of the experience. The museum visit provides a personal contact with the object of desire that is lacking in the consumption of the filmic Pygmy. In this context, the encounter with difference is magnified the closer one gets to the physical presence of the Other, an intensifying effect that reaches its apotheosis in the flesh and blood encounters of ethnic tourism. Such direct encounters are the sharp end of the acts of bulimic consumption. It is these face-to-face meetings that form the end point for the outward surge of myth, but also the beginning point for the next phase of the process. In this sense, all of the visual mediums referred to are both producer and product, stimulating and satisfying desire coterminously: when myth meets its mirror, it signifies the inevitability of the eternal return. It is at this moment that photographs and postcards become crucial aspects in the flux and reflux of myth. On the one hand, they trap the experience in time, capturing the minute in the mind of memory and crystallizing a particular collage of mythic associations. While on the other, they release the myth into the evanescence of the semiosphere, the act of consumption becoming of itself an act of production within the progressive circularity of myth.

Repetition

There is a broad pattern of repetition between and within the three levels that make up the matrix of *transmission/translation/reception* for the mythic formation of the Pygmies. They all interweave, confirming and transforming each other through patterns of reiteration that circulate within the mediaspheres of modernity. Within the idealization of the Pygmies, there is a continual layering of representation upon representation both horizontally and vertically: they slide in relation to one another and over one another in both space and time. Such imaginal repetition can be both simple and complex. In its simple form, it operates through a mechanical, stereotyped repetition of the same element, while complex repetition involves elements that multiply each other (Deleuze 1994). It is the interplay between these two forms that drives the myth, the propulsive energy coming from the repetitions of the material and of the spiritual. The regurgitation of the myth of the Pygmies relies upon the reproduction of physical and tangible manifestations of the Other and also upon metaphysical and abstract repetitions of desire.

Movement and transformation within these dynamics of myth can be identified through the ideas of pulsed and non-pulsed time (Deleuze and Guattari 1987). Pulsed time can be understood as the linear development of the myth of the Pygmies as it changes through time, whereas non-pulsed time is the random echoes of myth as they erupt from the shadows of the semiosphere. In general terms, there has been a progression within the myth that mirrors the colonial domestication and post-colonial appropriation of the Pygmies. The mythopoeia within the global semiosphere reflects the 'cyclical phase[s] in the world systemic of dynamic capitalism' (Friedman 2002: 33) and can be found refracted in the changing practices of tourism. As objects of the dominant gaze, the Pygmies have been transformed from savage exemplars into eco-saviours, from prototypes of the primitive mind into protectors of a forgotten wisdom, and from racist curiosity into relative wonder. And yet, within the courses and recourses of linear history, there are strange outbursts that break up the pattern and can even facilitate shifts within the aesthetic idiom. The smooth flow of myth is interrupted by incursions from the past and by eruptions from other mythic formations. These non-pulsed elements can be epistemic ruptures within the mainstream of pulsed time, as was the case with Schweinfurth's resuscitation of the *mythos* of the Ancients; or they can be merely blips on the surface static of the myth.

Figure 5.3   *Pigmies, Uganda-Congo*. Edition: 'Africa in Pictures'.
              Published by Pegasus Studio, Nairobi. Photographer S.
              Skulina, c. 1950-1959

Figure 5.4   *Pygmy Postcard* c. 1950. Publisher and photographer unknown

A further distinction within the historical trajectory of pulsed time can be identified within the formalized repetition of images. On this level of representation, we can distinguish between musematic and discursive types of repetition (cf. Middleton 1990). The musematic is the repetition of short units such as riffs or words, and, in the visual context, can be interpreted as the recurrence of tropes in and across the means of representation. For example, the same exoticization of the mundane that characterizes the filmic Pygmy is also apparent within the image repertoire of postcards. The commonplace activities of singing and dancing (Figure 5.3), and the everyday act of hunting (Figure 5.4) recur as regular images in both of these visual mediums as they do aesthetically within the textual and aural levels. Through the patterning that emerges from out of musematic repetition, an identifiable iconography of the postcarded Pygmy becomes apparent.

The 1906 card of *Colonel Harrison's African Pygmies* (Figure 5.2) incorporates many of the main 'accoutrements of the savage' that developed into a recognizable

Figure 5.5 *Pigmy***, Series No. HSM 88. Published by H.S.M. Limited, Mombasa, c. 1950**

Source: Author's collection.

Figure 5.6   *Pigmies, Uganda-Congo.* **Edition: "Africa in Pictures".
Published by Pegasus Studio, Nairobi. Photographer S.
Skulina, c. 1950-1959**

iconography of the Pygmy. Among this set of images are the props used to compose the portrait of the primitive such as weapons for hunting, musical instruments, and rudimentary clothing. Each of these items has become a visual synecdoche within more general recursive patterns. For example, the bow and arrow are the main element within the visual trope of hunting, a style most frequently replicated by the single shot of the hunter (Figure 5.5). This, in turn, recalls the pseudo-scientific photography of the nineteenth century, a style popularised by the postcard genre of the 'ethnic type'. In such pictures, the camera zooms in on the object, the close-up highlighting physical and physiognomic differences, principle markers of authenticity (Figure 5.6). That the body of the Pygmy is a prime focus for their exoticization is made clear by numerous postcards that include the tourist within the frame (Figure 5.7). In these, the presence of the tourist enhances the perception of bodily difference by providing the measure by which to judge the stature of the Pygmies. The contrast between the two has the further effect of accentuating the lack of clothing worn by the Pygmies. Their near-nakedness places them within

Figure 5.7   *Pigmies, Uganda-Congo.* **Edition: "Africa in Pictures".
Published by Pegasus Studio, Nairobi. Photographer S.
Skulina, c. 1950-1959**

nature, a method of situating that is often augmented by the appearance of the
rainforest or of their vernacular dwellings as a backdrop to the picture.

As Edwards has pointed out (1996: 213), once an image enters into the public
sphere, its meaning becomes 'free-floating' and open to the multiplicities of individual
interpretation. However, when seen as a collection, the combined elements that make
up the iconography of the postcarded Pygmy provide a visual grammar that helps us
to 'read' the individual images. This is supplemented by thematic recurrences that
form the semiotic background to the visualization of the mythic Pygmy. Underlying
the visual immediacy of the musematic repetition there is a pattern of discursive
repetition through which the aesthetic ideologies behind the image can be decoded.
It is ideas themselves that are repeated. In the postcards that contrast the statures
of tourist and Pygmy, the paternalist gaze of the ethnic tourist looks down upon
the childlike primitive. The sense of innocence this conveys upon the objects of
the gaze is embellished by placing them within a state of nature; however, there
is an ambivalence within this discursive repetition. Contained within these images

Figure 5.8   *Pygmée*, **Congo Belge 70. Published by Photo-Home,
                Leopoldville, c. 1950**

is a 'naturalizing hierarchy' (Kim 2002: 151) that recalls the civilizing mission of
colonialism and locates the Pygmies within an evolutionary chain of being. The
final postcard (Figure 5.8) in this chapter exemplifies this point. On the musematic
level, it repeats the fascination with the body of ethnic type genre, while on the
discursive, it reiterates the evolutionary bias that cast the Pygmies as the 'missing
link'. Beyond this, however, the beatific expression on the man's face resonates with
the eighteenth-century stereotype of the 'lazy, childlike' African as cherub (ibid.). In
one sense, this can be thought of as a non-pulsed echo from the past, but in another,
it can be seen as a continuation of the long history of racism that has underpinned our
engagement with the Other. As such, it is the inequality in the relationship between
the ethnic tourist and the ethnic object that is repeated within all of these images, and,
in this context, the politics of the postcard underlines the power contained within the
myth of the Pygmies.

Conclusion

Unlike Benjamin's (1969) notion of the loss of aura of the original, the reproduction
of the Pygmy began in simulation and has been transformed into an image that not

only holds onto its mythic, auratic qualities, but actually gains in them through reproduction. Each level and each singular representation has its own chronotope or timeline and can be understood in isolation. However, it is the combination of all of these aspects, the way that the prose, the picture, and the pop song come together, that sustains the myth. The intensity and durability of the particular brand of myth associated with the Pygmies, the vortical and progressive circularity of the bulimic consumption of strangeness, has locked them within an endless loop of hyperreality. Bishop (1989: 249), in his seminal text on the myth of Shangri-La, has put forward the idea that 'places echo' and that this echoing is one of the defining characteristics of a 'place's placeness'. In the same vein, it can be added that Pygmies also echo and that it is these polyphonic echoes, resonating across space and time, which suspend them within the dynamic stasis of myth. When these echoes are combined with the repetition of auratic reproduction, the memory traces of the 'travelscape' (cf. Tilley 1997) and the spatial memories of fear and desire (Bishop 1989: 249-250), the web of myth becomes so tightly enmeshed that no other thought seems possible. The endless and repetitive recombination of the mythologic, the intertwining of the *mythos* and the *logos*, overdetermines the symbolic configuration, but these overdeterminations become the expected norm for the people to whom they are applied.

It is through the modern methods of travel, through tourism, development, missionization and academic research that the myth of the Pygmies is returned to its imagined source. The repetition of image is matched by a repetition of desire in which the traveller consumes the symbolic as if it were real. It is only when the Pygmies that are consumed do not reflect the qualities of this myth back to the gaze of the spectator that the perception of aura is lost. It could be argued that reality presents itself when the symbolization is seen to be a failure or incomplete. However, this is not the case in relation to the bulimic consumption of Pygmies. Instead, the language of myth switches code; there is a movement from the language of romance that continues to reify Otherness to a language of extinction that blames the Self for the loss of the Other. In this nightmare scenario, the Pygmies are imagined as being stripped of a state of 'original affluence' (Sahlins 1968), reduced instead to a state of abject poverty. When this occurs, the Pygmies mutate from being a Rousseauesque fantasy into a Hobbesian nightmare, an equally potent mythic formation that allows for only one direction of development, the inevitable slide into cultural extinction. As a symbolic figure within the mythologic systems of the semiosphere, the Pygmies change from a sacred object into a sign of the wretched. In effect, a different yet equally powerful aura is attributed to them: the authenticity of the victim, the Other destroyed by the rapacious desire of the Self. This only reconfirms the need for the authentic Pygmy and the vanishing 'truth' of the eternal primitive. As the vanquished of modernity or as a valedictory of difference, the end result is the same. The Pygmies remain locked within the Pygmy paradox, the hyperreality of myth that judges every action against the fluidity of imagined ideals, with what they can and cannot be. There is a chain of antonomasia in which the mythic figure of the Pygmy refers to the general aesthetic of the primitive, which, in turn, stands for the ideology of authenticity. In the gaze of the cherub staring back at us, all that we can see is the reflection of our psyche and the absence at the core of our alienation. The Pygmy is no longer Other; they are the embodiment of our own fears and desires, the sense of loss deep within the Self. It is this lack that is regurgitated within the myth of the Pygmies, and it is this lack that continues to motivate the ethnic tourist.

References

Adams, J.S. and McShane, T.O. (1996), *The Myth of Wild Africa, Conservation Without Illusion*, University of California Press, Berkeley, CA.
Bahuchet, S. (1993), 'L'invention des Pygmées', *Cahiers d'études Africaines*, Vol. 129, No. XXXIII-I, pp. 153-181.
Barta, R. (1995), 'The Imperial Dilemma: Artificial wild men or supernatural devils?', *Critique of Anthropology*, Vol. 15, No. 3, pp. 219-247.
Barthes, R. (1973), *Mythologies*, translated by A. Lavers, Paladin, London.
Bauman, Z. (1993), *Postmodern Ethics*, Blackwell, London.
Bauman, Z. (1997), 'The Making and Unmaking of Strangers', in P. Werbner and T. Modood (eds), *Debating Cultural Hybridity: Multi-Cultural Identities and the Politics of Anti-Racism*, Zed Books, London and New Jersey, pp. 46-57.
Benjamin, W. (1969), *Illuminations*, edited by H. Arendt, translated by H. Zohn, Schocken Books, New York.
Bishop, P. (1989), *The Myth of Shangri-La: Tibet, Travel Writing, and the Western Creation of Sacred Landscape*, The Athlone Press, London.
Boorstin, D. (1964), *The Image: A Guide to Pseudo-Events in America*, Harper, New York.
Borges, J.L. and Guerrero, M. (1970), *The Book of Imaginary Beings*, translated by N. di Giovanni, Jonathan Cape, London.
Bradford, P.V. and Blume, H. (1992), *Ota Benga: The Pygmy in the Zoo*, St. Martin's Press, New York.
Bruner, E.M. (1991) 'Transformation of Self in Tourism', *Annals of Tourism Research*, Vol. 18, pp. 238-251.
Carey, J. (1992), *The Intellectuals and the Masses: Pride and Prejudice Among the Literary Intelligentsia, 1880-1939*, Faber & Faber, London.
Church, M. (2003), '*No Small Triumph* (Online document: http: //enjoyment. independent.co.uk/music/features/story.jsp?story=450438).
Clifford, J. (1997), *Routes: Travel and Translation in the Late Twentieth Century*, Harvard University Press, Cambridge, MA.
Crichton, M. (1983), *Congo*, Arrow, London.
de Quatrefages, A. (1895), *The Pygmies*, translated by F. Starr, Appleton, New York.
Deleuze, G. (1994), *Difference and Repetition*, translated by P. Patton, Columbia University Press, New York.

Deleuze, G. and Guattari, F. (1987), *A Thousand Plateaus: Capitalism and Schizophrenia, Vol. 2*, translated by B. Massumi, Zone Books, New York.

Edwards, E. (1996), 'Postcards: Greetings from Another World', in T. Selwyn (ed.) *The Tourist Image: Myths and Myth Making in Tourism*, John Wiley & Sons, Chichester, pp. 197-221.

Fabian, J. (1983), *Time and the Other: How Anthropology Makes its Object*, Columbia University Press, New York.

Feld, S. (1996), 'Pygmy POP: A Genealogy of Schizophonic Mimesis', *Yearbook for Traditional Music*, Vol. 28, pp. 1-35.

Frankland, S. (2001), 'Pygmic Tours', *African Study Monographs*, Supplementary Issue no. 26, pp. 237-256.

Frankland, S. (1999), 'Turnbull's Syndrome: Romantic Fascination in the Rain Forest', in K. Biesbrouck, Elders, S. and Rossel, G. (eds), *Central African Hunter-Gatherers in a Multidisciplinary Perspective: Challenging Elusiveness*, Univesiteit Leiden Research School for Asian, African and Amerindian Studies (CNWS), Leiden, pp. 61-72.

Friedman, J. (2002), 'From Roots to Routes: Tropes for Trippers', *Anthropological Theory*, Vol. 2, No. 1, pp. 21-36.

Green, P.J. (1999), 'A Revelation in Strange Humanity: Six Congo Pygmies in Britain, 1905-1907', in B. Lindfors (ed.), *Africans on Stage: Studies in Ethnological Show Business*, Indiana University Press, Bloomington and Indianapolis, pp. 156-187.

Grinker, R.R. (2000), *In the Arms of Africa: The Life of Colin M. Turnbull*, St. Martin's Press, New York.

Jahoda, G. (1999), *Images of Savages: Ancient Roots of Modern Prejudice in Western Culture*, Routledge, London and New York.

Kim, E. (2002), 'Race Sells: Racialized Trade Cards in 18th-Century Britain', *Journal of Material Culture*, Vol. 7, No. 2, pp. 137-165.

Lévi-Strauss, C. (1973), *Tristes Tropiques*, Jonathan Cape, London.

Lotman, Y.M. (1990), *Universe of the Mind: A Semiotic Theory of Culture*, translated by A. Shukman, Indiana University Press, Bloomington.

Lubbock, J. (1875), *The Origins of Civilization and the Primitive Condition of Man: Mental and Social Conditions of Savages*, Longmans, Green and Co, London.

MacCannell, D. (1992), *Empty Meeting Grounds: The Tourist Papers*, Routledge, London.

MacCannell, D. (1976), *The Tourist: A New Theory of the Leisure Class*, Schocken Books, New York.

Maffesoli, M. (1996), *The Time of the Tribes: The Decline of Individualism in Mass Society*, translated by D. Smith, Sage Publications, London.

Middleton, R. (1990), *Studying Popular Music*, Open University Press, Buckingham.

Rojek, C. (1997), 'Indexing, Dragging and the Social Construction of Tourist Sights', in C. Rojek and J. Urry (eds) *Touring Cultures: Transformations of Travel and Theory*, Routledge, London, pp. 52-74.

Rony, F.T. (1996), *The Third Eye: Race, Cinema, and Ethnographic Spectacle*, Duke University Press, Durham and London.

Root, D. (1996), *Cannibal Culture: Art, Appropriation, and the Commodification of Difference*, Westview Press, Oxford.

Rosaldo, R. (1982), 'Utter Savages of Scientific Value', in E. Leacock and R. Lee (eds), *Politics and History in Band Societies*, Cambridge University Press, Cambridge, pp. 309-25.

Sahlins, M. (1968), 'Notes on the Original Affluent Society', in R. Lee and I. DeVore (eds), *Man the Hunter*, Aldine, Chicago, pp. 85-89.

Schweinfurth, G. (1873), *The Heart of Africa: Three Years' Travel and Adventures in the Unexplored Regions of Central Africa from 1868 to 1871, Vol. 2*, translated by E. Frewer, Sampson Low, Marston, Low and Searle, London.

Shields, R. (1991), *Places on the Margins: Alternative Geographies of Modernity*, Routledge, London and New York.

Tilley, C. (1997), 'Performing Culture in the Global Village', *Critique of Anthropology*, Vol. 17, No. 1, pp. 67-89.

Tomas, D. (1996), *Transcultural Space and Transcultural Beings*, Westview Press, Oxford.

Turnbull, C. (1961), *The Forest People*, Chatto and Windus, London.

Urry, J. (1990), *The Tourist Gaze: Leisure and Travel in Contemporary Societies*, Sage Publications, London.

Van den Bergh, P. (1994), *The Quest for the Other: Ethnic Tourism in San Cristobal, Mexico*, University of Washington Press, Seattle.

Chapter 6
Photographing Race: The Discourse and Performance of Tourist Stereotypes

Elvi Whittaker

... the system of discourse by which the 'world' is divided, administered, plundered, by which humanity is thrust into textual pigeonholes, by which 'we' are 'human', 'they' are not... (Said 2001: 26).

The advent of photography in the nineteenth century gave added exposure to an existing categorization of human types that had prevailed for centuries. It made available to a viewing public a visual simplification of the knowledge that had developed colonialism and, in turn, had been further reified by it. The new graphic accessibility served to modify, convince, clarify, denounce, sensitize, question and accord credence to already existing beliefs. Photography was rapidly adopted as a recorder of truth and a conveyor of history. It also became one of the discourses of choice for the modern tourist industry, a visual text for defining other humans. The vivid images in photographs, the exotica that they promised were, in themselves, invitations to tourism. Thus tourism, photography and colonialism converged in the nineteenth century as they continued to do in the centuries that followed. While tourism and photography grew to be towering presences, colonialism came to be seriously undermined in many quarters, but it managed, nevertheless, to hang on as a deeply ingrained characteristic of Western sensibilities. Each of these three very complex discourses placed its own distinguishing stamp on a basic episteme. I refer to the enduring awareness of human differences and its unhappy legacy of race and racism. The notion of empire itself was clearly dependent for its existence on the politics of these differences. The role of tourism and photography in this particular relationship is much less obvious.

I direct my attention, in particular, to the explosive contribution made by tourist photography in the period 1880s to 1920s, to the racial drama of the last hundred years and to its role in establishing theoretical certainties about racial differences that have endured and are now considered detrimental to the human condition. It was the photography related to tourism that disseminated a pervasive visual discourse about race. It played an enduring part in creating a theatre of racial types, a gallery of visual stereotypes. These stereotypes were readily fueled by the prevailing nineteenth-century agenda of science which at that time was beset with the urgency to categorize human types, to fill in their anatomical and physiological report cards, to provide evidence for powerful theoretical scripts like that of evolution. In addition the stereotypes were important in social agendas such as cultural status, economic differences and, significantly, in colonial administration directing the prerogatives for interaction with colonized subjects – all crucial necessities upon which colonial empires were dependent. The stereotypes supported every notion of Western superiority.

The discourse on race is beset with monumental ethical complexity – the cause of an infinite number of social struggles over past decades. Widely criticized for its traditional and self-serving agenda, it is, however, on the eve of an authoritative rewriting. The authority for the reworking comes again from the powerful conglomerate of science and cannot be easily ignored. Science offers 'irrefutable evidence' in the findings on the human genome. Consequently centuries of mythologizing, pseudo-scientizing and politicizing about human differences could become nothing more than an evil Western spell. The analysis promises a significant readjustment of previous exhortations.

Consider the many theories employed in the past to make sense out of the array of human differences. What then is this phenomenon called 'race'? More importantly, what work was it destined to do? One could well question whether racial categorizing is anything more than a Linnean (and therefore scientific) attempt to bring order to apparent chaos by the introduction of usable categories resulting in the anchoring all human diversity into stable biological and social packages (Kroeber 1948)?[1] Is it a mythological system that acts as a defense against the fears produced by diversity and thus carefully engineered to permit domination to prevail? Stocking calls race a 'mythistorical archetypification process' (1991: 208). Is it a widely-accepted Christian ordering of human kind, as old as Noah and the migrations of his three sons over the earth, a biblical 'taming of the wilderness' and the establishing of dominion over lesser forms? Texts explicating the concept of race are frequently given to mentioning Ham, one of Noah's sons, who fell from grace and had his land and his sons cursed by Noah – the 'servant of servants shall he be unto his brethren' (Genesis 9:25). The descendents of Ham were believed to have peopled Africa. Perhaps race is not about biology and mythology at all but, as some would argue, about class and status, a human agenda imperially transported from Europe to the colonies (Cannadine 2001: ix). Alternately, is race just an early, an uncomfortable, but somehow necessary, stage in the inevitable process of global assimilation? Is it forever fixated as scientific support for physical

---

1   As anthropology played a strong hand in the early parts of the twentieth century in producing classifications of racial types, depicted most prominently in the well-disseminated textbook of Alfred Kroeber (1948). It is important to consider the significant change in the ethos of the discipline on this matter. By 1970 most members of the discipline had abandoned the analysis of race, or race as a concept: 78% used it in 1931, 36% in 1965 and 28% in 1996. Yet in thirteen textbooks between 1932-1969 where the existence of race was accepted and explored, in only one was its existence actually denied. In 1970-1979 twelve texts argued for its existence, fourteen argued against its existence. In 1990-1999 one text gave space to the existence of race as an analytic and nine were expressly against it (See Lieberman et. al., 2003).

evolution? Will this powerful set of beliefs, taken singly or together, endure beyond all attempts at social corrections? Is race a language, a *lingua franca,* a way to facilitate interaction in plural societies (Whittaker 1986: 141)? Is it merely an unresolved political position between the believers of the monogenetic versus the polygenetic origins of humans, that particular socio-intellectual war that consumed the American south in the nineteenth and twentieth centuries (Stanton 1960)? Thus is it a remnant of the religious tensions between beliefs that humans were descended from gods and fashioned in their image and the opposing scientific beliefs that humans ascended from the lowest life forms? Perhaps it is merely a system of exclusion, hiding its darker intent under a moral cloak of claimed visual evidence and objectivity? Race is, of course, all of these things and more. Part imagination, part 'necessary' metaphor, part 'motivating myth', part 'subversive enigma'. Despite the swelling opposition, 'race' is assumed, nevertheless, to have significance in what it indicates, all too clearly, about the existence of a deeper truth about those who proclaim and support it.

Establishing the Visual Stereotypes

At the time when there were naïve beliefs about classification, about recording human differences and rendering them scientific, photography became available to facilitate the process. Then tourism, what van den Berghe calls 'a special form of ethnic relations' (1984), played its part and disseminated the results. It was timely, occurring when colonizers in Africa, India, the Pacific, and in the Americas, as well as the tourists who came in their wake, sought guidance in their interactions with Indigenous people, especially craving explanations that went beyond brute force and complete subjugation. Given the deficiencies of understanding current at the time, colonizers perceived what they called lethargy and apparent lack of intelligence on the part of colonized natives. This they explained by 'racial' differences. Efforts at understanding were bolstered by European intellectual systems developed in the Enlightenment – an all-powerful positivism that supported discrete categories on the one hand, tempered by an equally persuasive romanticism proclaiming sentiment and humanism on the other.

Before the emergence of photography the general population of Europe and North America had limited knowledge and sparse visual imagery about people in the colonies in Africa, Asia, the Pacific, and North America. Descriptions of these people, imaginary and biased as they were, appeared in popular literature such as the novels of H. Rider Haggard, Karl May and Edgar Rice Burroughs. Visual imagery was even more elusive. Pictorial depictions seldom appeared in newspapers and the exposure to periodicals like *The London Illustrated News* or *Harpers' Weekly*, where traveling artists supplied illustrations, was limited. The educated fared a little better having access to some of the illustrated works of early anthropology, to various atlases of races and to encyclopedic works like Ridpath's *The Great Races of Mankind* (1893). The latter offered accounts of the origins, early migrations, social evolution, 'primitive estate', present conditions and the future promise of the principal families of 'man'. Anthropology in general has had a long history with photography, in recording ethnographic scenes as well as in contributing to an ongoing classification and documentation of races (e.g. Edwards 1992; 2001).

Meanwhile the tourist encountered 'the racial other' in the colonies, and acquired photographs and postcards of such exotic inhabitants. These depictions became part of the trophies collected in distant lands to be sent home as mementos of the traveler's experiences. Or, alternately, they were purchased at home as postcards and became part of the photographic collections of the bourgeoisie of Europe and America. If not received from family and friends either as a means of communication or as contributions to collections of such images, these cards were bought from vendors of every kind, in shops, on the streets, at train stations and at parks and other public recreational sites. Neither senders nor receivers were necessarily travelers. Such opportunities to view photographs of peoples around the world edified a populace who never ventured outside their own country, and often not beyond their own county, city or village. Commonly known as 'armchair travelers', this large population, then as now, lived in an imagination fueled by literature, folklore, newspapers, traveler's tales and photography. They thereby avoided the ego-threats of high expenses, packing, foreign languages, lost luggage, challenging or nonexistent maps, excessive heat or cold, sickness, thievery, ridicule, unforeseen costs, uncomfortable or unobtainable transportation, shouldering strangers, threatening diseases, bad beds, poor food, nonexistent and inadequate toilet facilities, accidents and other mishaps. Yet they are tourists of a sort just the same, for the culture of touring does not only involve a geographical displacement.[2]

Tourist photographs and postcards, whether actively acquired in the pursuit of adventure or in a sedentary touristic culture, have ensured a dispersal of images from the far corners of the world. Millions of photographs passed through the mails. In Britain alone, in the year 1909-1910, 866 million postcards found their way through the post. The post arrived three times each day, and a message sent in the morning could have a response by afternoon. Statistically, this was an average of twenty postcards a year for every individual in Britain (Staff 1966; Whittaker 2000).

---

2   These travelers of imagination are largely ignored by the academic arm of the tourism industry. Yet they are targeted continually by the industry itself in the hope of luring them into the wider geographical and commercial world of travel. The perpetually-prospective traveler deserves recognition, if only for their 'expectations', a concept which has long been a preoccupation for academics. Thus such voyageurs of imagination, if action is unrealized, remain silent consumers and partners in the ever-growing literature on travelers' tales, travel photographs and other travel related products.

The Conventions of Performing Race: The Grammar of the Photograph

Tourist photographs reflect a visual literacy about the reigning theories of race. They are complicated photographic discourses that engage other discourses beyond themselves (Burgin 2001: 75) as latent texts of all kinds crowd into the play of meanings. The claims of photographic neutrality, of images presenting things as they are, was pervasive in the nineteenth and the early twentieth century but can no longer be taken seriously. The certitude that photographs mirror nature, that they offer 'reality' unadulterated has long been overturned. Yet the lingering existence of the discourse proclaiming exact duplications of reality and nature give photographs a 'representational legitimacy' (Hamilton 1997), a powerful presence.

Despite all claims to the contrary, photographs are hardly neutral. The 'taking', or having a photograph 'taken' is itself a kind of victimization, a rite of dispossession, not unlike the taking of territory. It involves assuming the right to control representation and the text that ensues. In the early days, images were appropriated often without consent. Indeed it would not be amiss to view every photograph as an act of aggression, however benign or well intentioned it may be. Sontag suggests that a photograph is a 'tool of power' (1977: 8) and Corbey that 'taking pictures was indeed another means of taking possession of native peoples and their lands' (1993). To photograph the colonized person is to colonize that person a second time. Not surprisingly, the faces of those depicted look directly into the camera. They are invariably grave, vacant in expression, bewildered, a little defiant. The technology of the camera of course dictated some of this motionless gravity and a happy demeanor is exceptional.

Early photographs of colonial subjects were carefully staged as, indeed, were all photographs. Each rendering raised the question as to whether it was taken originally by a tourist, a government official, a missionary, a military man, professional photographer with a studio or a visiting freelance photographer hoping to sell to postcard companies. The most unlikely possibility was that it was the creation of a native entrepreneur.[3] What is not clear is the nature of the interaction between the photographer, by definition a Westerner, and the colonial subject. Some mediation undoubtedly occurred to produce the images, whether it was for administrative purposes or for commercial use. What kind of intrusion was it, under what conditions? What persuasions would have influenced individuals with strong inclinations to protect their privacy or, in some cases, the fear of their souls being captured? It is not credible that subjects would have asked for their portraits to be recorded. Did a financial transaction occur? What was the nature of coaching for positions and expressions? What did the photographer intend for the finished product? Did they sell images directly to visiting tourists, businessmen,

---

3 Indigenous photographers did exist but, unfortunately, their work did not appear in the literature. Among the few exceptions is George Hunt whose photographs of Kwagiutl life are an important part of the record of this Northwest Coast tribe (Jacknis 1992).

Figure 6.1   *Group Otago Maoris* **(1908) photographed by Gill. F.T. Series
No. 2339**

government officials, anthropologists or postcard companies for redistribution?
What cultural prerogatives influenced the photographer? The camera-wielding
activities are not known. Innocent subjects had their private spaces invaded and
became forcibly enmeshed in a discourse beyond their control. Photographers
produced Indigenous portraits, but even more importantly, they produced a
revealing portraiture of those behind the camera. This world behind the camera is
even more visible in the repertoire of stereotypes evoked in the process.

The performances for the camera are generally not considered in the analysis of
photographs. Theatricality and narrativity characterized the performances for most
early photographs (Pauli 2006). The cultural reading of the photographs, however,
always assumed the depiction of reality and natural life, not fiction. When apprized
of the performance and the staging the viewer often expressed disbelief and felt
deceived. The expectation of reality and the natural in photographs continues
into the present century, enforced by decades of photography as an everyday and
widely-shared pursuit, where the aim is to 'capture reality'. At the same time the
professional photographer in the artistic world has moved wholeheartedly into

narrative, performing, and constructing scenes as seen in the work of Jeff Wall, Andreas Gursky, Gregory Crewdson and Cindy Sherman among others.

The legendary photographer of American tribal people, Edward Curtis provides some information about the staging of his photographs. In the pursuit of 'authenticity', he dressed subjects with wigs and used costumes and decorations from his own collection of aboriginal ceremonial items (Gidley 1992). Similarly, Maxwell (1999) in her study of exhibition photography describes the positioning of subjects for what was considered an appropriate scene. A photograph of a Maori tribal group in 1907 included Professor (Sir) Peter Buck (of anthropological fame, a physician and a member of the New Zealand parliament), Captain Gilbert Mair, who had led the Maori in the Maori Wars, and Maggie Papakura, a renowned tourist guide who later married a British earl and studied at Oxford. To appear 'authentic' as traditional Maori, these three essentially urbanized people wore appropriate tribal clothing and were positioned in front of a traditional village, flanked by armed Maori warriors (1999: 137-38). Figure 6.1 depicts a photograph of a Maori group in the early years of the twentieth century, sent as a postcard from New Zealand to London in 1908. It assumes the positioning and stance of family photographs familiar to all Western people, with an innocuous background, some members standing behind others who are seated, all looking directly at the photographer. Each holds or wears articles declaring a cultural affiliation.

The urbanized status of the group is clearly visible in the trimmed facial hair of the males and the styled haircuts of the women. As with all such early photographs, the names of the individuals remain hidden, apparently unimportant, in this recording of cultural accoutrements and racial visages.[4]

Photographers in colonial cultures were actively engaged in the newly lucrative undertaking of photographing Indigenous people. Some cast a practiced eye on passers-by, jumping into action upon sighting a prospective subject and inducing that person, perhaps with flattery, force or cash, into their studio to sit for a portrait. Initially such captured subjects were positioned in exact replications of European photographic and painted portraiture. The full frontal view, which seems to have been the norm, takes on the severity of the mug shot, the scientific illustration, the passport photograph – and, like these documents, indicates an institutional incorporation. Occasionally, the subject was asked to place a hand on some prop

---

4 After a public lecture in Vancouver where I had projected this photograph, I was approached by a man who confessed to being surprised to recognize a Maori grandparent. He regretted the anonymity and loss to history of such early public photograph. The ethics of Indigenous photography have changed significantly during the last hundred years and there is a growing trend, especially in the last decades, to attach names. For some of the older photographs disseminated as postcards and in the public domain for some time, the addition of a name such as Captain Joe, or other essentially denigrating colonial reference, under the guise of honouring the subject, pays no actual respect. An exception to the no-naming convention in early tourist photography is the case of the photographs of North American First Nations' chiefs, who are often identified by their English names.

like an artificial tree-trunk, a railing or the back of a chair for steadiness during the long procedure. Sometimes the subject appeared before a backdrop with a romantic theme. In using this conventional positioning the Indigenous photograph followed the photographic literacy of the day.

The Rituals of Abstraction

Subjects recorded on film have a strange reality bestowed upon them, as part of the photo-politics and tourism politics of the day. They are given a new permanence, suspended into the completely unnatural stillness possible only in photography. An irrevocable transmutation takes place. They are taken from the moving reality they occupy in life in which their full humanity is recognized, and fixed into abstractions of themselves. Motionless as specimens. Motionless as death. Not only does this provide a valuable service to programs of classification but also, in doing so, constitutes 'a ready iconography of the very process of civilization itself' with a fixedness that is 'an important feature of colonial discourse, its dependence on the concept of "fixity" in the ideological construction of otherness' (Bhabha 1983: 18). The power to make the classification, to produce the code, is in the hands of the photographers, servants of the prevailing Western ethos of evolution. The graphic imagery and the knowledge it supports can now be transported worldwide, or more directly into the offices of the colonial administrator, as 'specimens', meant to be displayed, analysed, documented, filed, added to collections of exotica, or to cabinets of curiosity. While the person photographed may well be encouraged to feel honoured to be thus represented, they have no knowledge of the kind of privileged cultural discourses that their image is used to support. Their powerlessness and inferiority is thus further confirmed (Said 2001). Their photographs contribute to institutionalized exclusion, the 'them' and the 'us' become even more firmly established (see also Hall 1997). These are not ultimate truths, even if such a thing as an ultimate truth could exist, but rather visual fictions passed on as facts. It is a familiar but apt observation that cameras record images from both sides, in front of it and behind it. A photograph is a display of the culture, the mind-set of the photographer, the ethnocentrism and the resulting 'visual decisions' (Sontag 1977: 89). The culture behind the camera calls for explication.

There is a stoppage in time and place, a formalized reductionism, an unspoken commitment to essence, an utter simplification, a facile demystification and the extraction from all cultural references and this is passed on as facts about 'race'. It arises out of white imagination and, in its turn, bolsters it. The received facts justify already existing 'factual' information. The lowly status of the person as specimen, without vestiges of privilege, is reaffirmed. A patronizing decision has made 'race' part of the colonial vernacular. Stocking (1991: 68) has called this a 'shadowy dance of archetypes from the dreamtime of anthropology'.

Photographic images of racial stereotypes have endured through generations of change in racial politics and in tourism. They have become symbols of the very essence of being Indian, Japanese, Inuit, Welsh, Aboriginal, Saami, Mexican, Hawai'ian and so on – their longevity assured through continuing usage. The imagery transcends the actual portrait, the actual individual and the actual time to become a certified discourse about the 'race'. It outlives its ephemeral object. A single image, a single face stood for countless others. Picasso is said to have observed to Gertrude Stein that 'no one will see the picture, they will see the legend of the picture. It makes no difference if the picture lasts or doesn't last'. Such stereotypes are necessary in cultures where race and ethnicity have long endured as ways of organizing the world, of guiding everyday interactions.

A series of French photographic postcards issued in the first decades of the twentieth century, entitled *Scènes et Types,* captured the ethos of stereotypes almost as if they were part of an anthropological or philosophical treatise. The images were numerically catalogued in the tens of thousands, representing most known geographic ethnicities. They proceeded from the simple recordings of the major races by skin colour – *Les types noirs, les types rouges, les types jaune,* where the locations and cultures were omitted as if such affiliations were unimportant. They progressed to the documentation of racial types from specific geographical locations, using ethnic clothing as identifiers such as *Types de Caucase* (N. 29), *Arabe de la Plaine* (No. 6258). Sometimes they moved to greater specificity by adding the occupation of the individual(s) to the prerequisite traditional image, for example fortune-telling, *Guidzane-Diseuse de Bonne-Aventure* (No. 2006).

Beyond the concerted 'scientific' efforts of the company of French publishers and photographers, simple stereotyping was the preferred depiction in the early years of tourist photography. The Scots appeared in kilts and as pipers, Hawai'ians were represented usually by women dressed for the hula, with grass skirts, leis and ukuleles, American First Nations were chiefs in tribal feathered headdresses, African Americans were pictured in the presence of watermelons or cotton, Middle Easterners had camels as their props, Australian Aboriginal males held boomerangs and spears, the Dutch wore clogs with backgrounds of windmills or tulips – in short, entirely predictable even a hundred years later.

An early but prevailing image of the Mexican is presented in Figure 6.2 – indolent, taking a siesta, evading responsibility. If this stereotype did not produce the immediate recognition expected in the viewer, the postcard company moralized on the back of the card with a proclamation: 'If one has absolutely nothing to do, or suffers from the constitutional ailment of having been born tired, Mexico is the place for him to rest. There is nothing more pronounced than that of procrastination. One of the native customs is to doze for a couple of hours after lunch and get busy as the sun nears the horizon. In this view is shown one of the common, everyday type of peon taking his siesta on a public thoroughfare'. This image was widely disseminated in the United States and reproduced in ceramics, toys, tourist advertisements, store signs and so on. While immediately recognizable such stereotypic depictions in no way embodied the vast body of

2866.   The Noon-Day Siesta, Mexico.

Figure 6.2   *The Noon Hour Siesta, Mexico.* **Photographic postcard, H.H.T**
                   Co. No. 2866

photography of the Mexican people produced in this period. Seldom seen were
images of pressing social relevance, such as Mexican poverty, schools, urban
and rural households, the life of soldiers and labourers, farming, industry and an
infinite number of other subjects. Some of these actually became photographic
postcards and appear now as historically interesting oddities, seldom associated
with the Mexico of the tourist mind. Photographs of the Mexican revolution and of
American military involvement are examples of the imagery that took the form of
postcards (Vanderwood and Samponaro 1988). The actual dissemination of these
powerful photographs, however, is minimal.[5]

---

5 Such photographs indicate the confusion about which photographs were the
work of amateurs, which were the work of professionals and which were intended as
commercial postcards. This confusion was exacerbated as for some decades both amateur
and professional photographs were printed on photographic paper the weight and size of
postcards and additionally imprinted on the reverse side with the postcard schema for

43. RASTUS AND NED.

Figure 6.3   *Rastus and Ned.* (c. early 1930s) C. T. American Art, No. 43, #
1914, published in Chicago

The demeaning myth about the love African Americans are supposed to have
for watermelon has spread from the American South, where it first appeared, to
elsewhere in the United States and beyond to other sites of African American
culture such as the Bahamas and the Caribbean. Figure 6.3, a photograph from the
early 1930s, bears witness to this belief and, under the attempted guise of humour,
manages to be persuasively demeaning. Usually such watermelon-eating scenes
depict young boys and occasionally adult males. As a rule women and girls are not
photographed with watermelons. Like many such photographs this one appears
to be orchestrated, capturing a completely joyless expression on at least one face.
Frequently the eaters stare resentfully at the photographer as if in protest against

address and message. This practice seems to have emerged in World War I, permitting
speedy delivery at that time and was extensively used until the 1950s. I am indebted to Brian
Stokoe and Alison Nordstrom for information on the printing of early studio photography.
See also <http: //www.payle.com/postcards.html#realphoto>

**Figure 6.4   Untitled Photograph. Presumed Hawai'ian with ukulele
(c. 1920s). Private collection**

the pose they are asked to assume. There are also numerous images of African
Americans at labour in cotton fields, stacking bales or sitting in the midst of the
downy material. Life on plantations, the play of children, the preoccupations of
the aged and other similar scenes of African American life are seldom seen. A
denigrating plethora of cartoon and comic postcards took their place and flooded
the industry for many decades living on in imagination and memory (Mellinger
1992).

The Hawai'ian stereotype is fully recognizable in a photograph from the 1920s
(Figure 6.4). It is taken in what appears to be at a local tourist site and the grass-
skirt, the ukulele and the lei are destined to suggest a hula dancer. Interestingly,
so strong are the trappings of the abstractions of Hawai'ianess that the viewer is
liable to overlook questions about ethnicity and gender that the photograph could
suggest.

The photograph mirrors another trend in Hawai'ian tourist photography. So
vigorous were the objections of the Indigenous Hawai'ian population to American
annexation in 1898 that a political decision was made to produce positive and
beautiful images about the islands (see also Maxwell 1999). The Hawai'ian beauty,

the hula dancer, was perhaps the most ubiquitous result. The bare-breastedness that continued much longer with other populations in the Pacific and Africa disappeared from Hawai'ian public photography by the 1920s. All too frequently, and perhaps teasingly psychoanalytic, such exposure was occasionally replaced by devices such as two pineapples held over the breasts. In fact, so classic is the Hawai'ian female stereotype, breasts bare or covered, that it was widely adapted to velvet paintings, tourist souvenirs and even disseminated to racial and ethnic minorities elsewhere. It seems not to have mattered in the production of this ideological subject that the women are increasingly less and less Polynesian/Asian looking and more and more *hapa haole* (half-white), or even clearly white (Maxwell 1999: 218).

Two particular portrayals became widely favoured by photographers for depicting women in the colonies. The male-centredness of these interpretations became pointedly obvious during the rapidly changing ideologies of the last quarter of the twentieth century. It is particularly telling that these conventions were widely used to penetrate the mysteries of women from the Arab world, those mythic creatures of supposed licentiousness, sexual recklessness and extravagance with sexual favours. Mythic images of odalisques were liberally borrowed from painters of earlier centuries and appeared in photography which captured female images in the non-Western world (Whittaker 2000: 426). The first portrayal depicted the imagined freedoms that were believed to exist in Turkish seraglios, where the *odaliks* (literally bed companions, female slaves or concubines) are believed to bestow their favours. Though the European imaginings of these legendary places were far from the mark, the imagery continued almost as a pornographic idea in the nineteenth century, and was liberally embellished and merged with the subterranean threats of 'the white slave trade'. The *odalisque* itself was a depiction favoured by Delacroix, Ingres, David, Renoir, Manet, Gauguin and Matisse (Board 1992). As the theatricality is so obvious in each photograph, the act of de-robing the subject almost seems part of the savoured experience of the act of photography itself, to say nothing of the experience of the viewer. The European symbolism was exported from the Arab world to south Asia as seen in Figure 6.5

The posing of the Ceylonese dancing girl on a photographer's couch with a backdrop of other *mis-en-scène* articles is typical of *odalisque* portraiture. The young woman is scantily clad, hardly the clothing the typical Ceylonese woman would wear on the street or elsewhere. It has the appearance of being added onto the print as an after-thought, draped softly with gossamer thinness over a seemingly naked body to reveal just the right amount of forbidden flesh. One breast is bare, much in the tradition of early religious paintings where a bare breast of the Madonna is 'legitimately' exposed and meant to reflect the nurturing mother, often complete with a suckling Christ child (Miles 1992). The leg placed on the couch does not immediately convey the expected feminine gracefulness. From this carefully constructed pose the subject stares confidently and unblinkingly at the viewer with an expression of mild amusement. Such seductiveness was in complete contradiction to the photography of typical European and American women and had more in common with the infamous Parisian postcard of the Victorian era.

Figure 6.5   *Nautch Girl, Ceylon* **(c. 1908). A photograph, numbered 7, produced and distributed by John & Co., Colombo and printed in England**

The further portrayal of women in the colonies is the photography that captures them fully nude above the waist. This preoccupation with mammary development and the label of sexual availability it suggests continued for many years as the recognizable stereotypic characterization of colonized women, assuming the same level of referential significance as the more usual native clothing or significant cultural prop. These revelations of the body, induced and staged in the privacy of the studio, create a gendered stereotype for later public viewing, possibly worldwide if converted to a postcard. Such missives could have numbered in the millions.[6]

At the time the images were captured, such depictions did not occur publicly nor were even permissible in the cultures from which they were obtained. Alloula (1986) writes of the proliferation of photographs of bare-topped Moorish women in the first decades of the twentieth century. His analysis reveals how French colonial politics merged with the male politics of the French military to transform completely the traditional religious and cultural demands which closeted Algerian women from the eyes of strangers. They were created as bold-faced, semi-nude temptresses. Such commodification of eroticism, whether producing photographs for private sale, or for dissemination by large postcard-producing companies,

6   The popularity of these cards endures to the present. In recent months an auction on eBay advertised, for a considerable sum, the sale of 500 old postcard photographs of 'bare-breasted native women

provides a clear example of how sexually exploitative images flooded tourist cultures in Europe and elsewhere. Licentious images such as these have come to represent the ever-present preoccupation with gender in the racial 'other'.

Performing the Inferiorities of Status

Western colonialism and capitalism had a presence over much of the world. The appropriation by photography was merely one of the many symbols of this power and among its multiple justifications in the nineteenth and early twentieth centuries was that Indigenous people were members of 'dying races' and doomed cultures. The imperative to record them, therefore, was self-evident. The ruling social philosophy of the time proclaimed that assimilation was the only avenue open to cultures under threat of extinction. This view mediated all interactions between the dominant Europeans and Americans and colonized races. Extinction was seen as particularly imminent in races deemed as being in the evolutionary stage of pre-literate 'savagery', itself a category bestowed by science. The other stages were 'barbarity' and at the peak of evolutionary progress, 'industrial civilization'. The debased category of 'savagery' was perceived to be true of beggars in European cities, those caught in the demise of once great civilizations, as well as in the conditions of 'primitive' peoples. As Ridpath points out in his four-volume *Great Races of Mankind* (Vol I, 1893), people 'without social and civil institutions', who practice cannibalism, polyandry or polygamy, or who are for one reason or another 'examples of the lowest stages of human development' belong to this group. Scientists and the public alike were convinced that the inhabitants of the colonies in the nineteenth century were in the process of cultural and physical deterioration. This view endured well into the following century and in different forms even into the present. Voluminous photographic records were created to support the classifications, often assuming postcard status and traveling through the mails. There were graphic depictions of the Indigenous people of North America, Australia, Africa and elsewhere in obviously degraded circumstances. There were images of pathetic attempts to adopt Western clothing, and adapt to colonial pursuits. Photographs depicted degeneration of various kinds, such as African Americans scrambling for coins thrown in their midst, dejected First Nations' families sitting on the sidewalks of city streets, Mexicans hunting for lice in each other's hair.[7] The moral message was the urgent need for assimilation and thereby a modicum of refinement. In these apparently degraded circumstances

---

7 Pointedly demeaning images were eliminated from this chapter as some faces are of excellent photographic clarity and easily recognizable as known persons. It is also important to indicate that some attempts are underway by Native American Indians, particularly the Hopi, to establish intellectual property rights over photographs in archives, museums and public collections, especially those that deal with sacred, religious and ceremonial objects and events (Downey 1998).

missionaries developed their own cadre of photographs and a significant number of postcards produced particularly to be sent home to encourage supporters that the work of the Lord, while obviously challenging was being successfully pursued. Such depictions contrast sharply with the 'noble savage' images of feather-bonneted chiefs and great warrior braves widely popular and claiming stereotypic status for North American First Nations people, or for the elite of other non-western cultures.

It has frequently been argued, and the photographs of degradation support the claim, that the relationship between the colonizers and colonized subjects, as well as Indigenous people everywhere is not one of race but rather one of class and socio-economic status. The class system of Europe was transferred quite successfully to the social soil of the Middle East, Africa, the Americas, India, Asia, Australia and the Pacific by the bourgeois administrators dispatched to bring order. The working class poor, the servant class poor, even the 'undeserving poor', so plentiful in European countries, were readily rediscovered among the Indigenous people elsewhere (Cannadine 2001: 8).[8] This class difference allowed the colonial system to work. As if playing out the proper scenario on film, Figure 6.6 shows a light-hearted performance for a tourist photograph. The scene reveals two Englishmen, possibly tourists or commercial travelers, posing leisurely in reclining chairs. The back of the photograph informs us that it was taken at Watsons Hotel, Bombay, on March 10, 1894. On the front is inscribed 'A talk about Ladies', below an appropriately constructed scene, suggesting that the photograph was to be sent back to England to signify that home was in their thoughts. As if to ensure that this theme be clearly understood, one holds in his hand a picture of 'a Lady'. What is also interesting, however, is that an Indian servant is pointedly positioned in the background, standing attentively, ready to do the gentlemen's bidding. Subservient and, unlike his masters, clearly not amused.

Perhaps the most derogatory claim to superior status was in the imagery of capitalism offered to those at home. The economic benefits extracted from the colonies is well documented and need not be exhorted here. Photographs showing the orderliness of production and the fortuitously happy deployment of colonial labour were highly praised a hundred years ago as good examples of the hard work ensuing in the colonies, of the cooperation of the native work-force and, most rewardingly for sender and receiver alike, the promise of prosperity. These same images today, in different cultural times, become examples of Western exploitation. Many major multinational corporations of the twenty-first century have embarrassing colonial pasts carefully closeted in archives and seldom seen. Figure 6.7 is a postcard produced by the Lipton Tea Company, probably in

---

8   Photographic depictions of the working poor made in the 1920s by a professional photographers like E.O. Hoppé constitute 'anthropological' documentation of everyday occupations. Yet the elegant skill that placed these photographs in the record avoided any stereotypical configurations, and the images remain as works of art, defying statements about class and status. For a discussion of Hoppé on *London Types* see Baggott and Stokoe (2003).

Figure 6.6   *Watsons Hotel, Bombay* (1894). Tourist photograph,
photographer unknown

great numbers, to be sent far and wide as advertising for the company. As shown
the company is excessively proud of its record and the resulting *trompe l'oeil*
photograph is an example of what Dubin (1990) calls 'visual onomatopoeia',
operating on a close equivalence between a subject and its representation, a 'living
photograph' destined to mobilize allegiance to capitalistic exploits.

The raw power that was able to command the above tableau cannot be
mistaken. The large numbers of 'coolies', at least 170-strong, involved in this
demeaning exercise broadcasts omnipotent control and military-like organization
of the work force.

At the same time as such photographs were taken, the privileged Europeans
consorted with the upper classes and the elite of the colonies. The sordid white
settlers, the lower class workers and the petit bourgeois administrative clerks,
who accompanied leading officials from the home country to establish the
national presence overseas, remained far separated from the power of government
administration and the management of the great companies. The déclassé Europeans
remained in their lowly slots. The European and American elite shot tigers with the
Maharajahs, feasted with the kings of African states, married Hawai'ian royalty, rode
with Arab Sheiks, writing about them as if they were English gentlemen. In India
the British introduced three new orders of chivalry solely for prospective Indian

Figure 6.7   *Muster of Coolies: Monerakande Estate Ceylon.* (c. 1910)
Lipton Series, published by C.W. Faulkner and Co. Letd
(sic)., London, EC. Printed in England. It is addressed to
Staffordshire

recipients. Clearly the social distance between the elite ruling class in the colonies
and the lower classes brought there from their home country was far greater than
that between themselves and the elite in native societies. The colonial photographic
catalogue is full of exotic sheiks, ruling emirs, native princes, Polynesian kings,
powerful chiefs as well as women of high status. The colonizers recognized and
continued to enrich privilege wherever they found it. In contrast to the pictures of
degradation, there are a myriad of Rousseau images, idealized, romantic, handsome,
resplendent in nobility and self-importance (see also Tamplin 1992).

In the last couple of decades the colonial discourse has been reworked by
movements of resistance in all parts of the Indigenous world. 'Race' has been
openly declared 'a significant marker of positioning and power' (Harrison 2002:
145) and thus deemed outside the pale of contemporary politics and policy
(Whittaker 1999). Race and class have become an embarrassment to the tourist
world, even though the prospects of viewing and experiencing Aboriginal/Native/
First Nations cultures remain an unabated draw for the average tourist. Postcard
depictions of anonymous actual persons have declined and in some places stopped
entirely. Sometimes the gap between the apparent preoccupation that the tourist is
presumed to have to 'tour' these cultures and the emerging discourses about the
appropriate treatment of 'the other' is being filled with substitutes of contemporary,
politically appropriate, photographic postcards and by the reproduction of carefully

chosen archival photographs. Thus, these latter depictions are viewed in the political present while having occurred in the supposed romantic past. The viewer merges these images with the current post-colonial discourse. Formerly colonized communities are producing their own original postcards and photographs, politically structured, ostensibly to educate tourists and change perceptions. These images are often visible corrections and critiques of the crass photographs of the past. Clearly, among the emerging ethics of the tourism industry, considerable attention is being conferred on the visual record.

Acknowledgements

This chapter is taken partially from an earlier paper read at the Annual Meeting of the Popular Culture Association, San Diego, California, April 1999 and also from a paper presented at the International Conference on Tourism and Photography in Sheffield, July 2003. Gratitude is owed to Lelia Morey, Neil Eaton, Mike Robinson, Brian Stokoe and Carl Whittaker for various kinds of assistance.

References

Alloula, M. (1986), *The Colonial Harem* (Theory and History of Literature, Vol. 21), University of Minnesota Press, Minneapolis.

Baggott, S.-A. and Stokoe, B. (2003), 'The Success of a Photographer: Culture, Commerce, and Ideology in the Work of E.O. Hoppé', *Oxford Art Journal*, Vol. 26, pp. 23-46.

Bhabha, H. (1983), 'The Other Question: the Stereotype and Colonial Discourse', *Screen*, Vol. 24.

Board, M.L. (1992), 'Constructing Myths and Ideologies in Matisse's Odalisques', in N. Broude and M.D. Garrard (eds), *The Expanding Discourse: Feminism and Art History*, HarperCollins, New York, pp. 359-379.

Burgin, V. (2001), 'Looking at Photographs', in M. Alvarado, E. Buscombe, and R. Collins (eds), *Representation and Photography: A Screen Education Reader*, Palgrave, New York, pp. 65-75.

Cannadine, D. (2001), *Ornamentalism: How the British Saw Their Empire*, Oxford University Press, Oxford.

Corbey, R. (1993), 'Ethnographic Showcases 1870-1930', *Cultural Anthropology*, Vol. 8, pp. 338-369.

Downey, L. (1998), 'A Tourist Album and its Implications for the Intellectual Property Rights of Indigenous People', *WAAC Newsletter*, Vol. 20, pp. 1-8.

Dubin, S.C. (1990), 'Visual Onomatopoeia', *Symbolic Interaction*, Vol. 13, pp. 185-216.

Edwards, E. (ed.) (1992), *Anthropology and Photography 1860-1920*, Yale University Press, New Haven.

Edwards, E. (2001), *Raw Histories: Photographs, Anthropology and Museums*, Berg, Oxford.

Gidley, M. (1992), 'Edward S. Curtis' Indian Photographs: A National Enterprise', in M. Gidley (ed.), *Representing Others: White Views of Indigenous People*, University of Exeter Press, Exeter, pp. 103-119.

Hall, S. (1997), 'The Spectacle of the "Other"', in S. Hall (ed), *Representation: Cultural Representations and Signifying Practices,* Sage, London, pp. 223-290.

Hamilton, P. (1997), 'Representing the Social: France and Frenchness in Post-War Humanist Photography', in S. Hall (ed.), *Representation: Cultural Representations Signifying Practices*, Sage, London, pp. 75-150.

Harrison, F.V. (2002), 'Unraveling "Race" for the Twenty-First Century', in J. MacLancy (ed.), *Exotic No More*, University of Chicago Press, Chicago, pp. 145-166.

Jacknis, I. (1992), 'George Hunt, Kwakiutl Photographer', in E. Edwards (ed.), *Anthropology and Photography 1860-1920*, Yale University Press, New Haven, pp. 143-151.

Kroeber, A. (1948), *Anthropology*, Harcourt, Brace, New York.

Lieberman, L., Kirk, R.C. and Littlefield, A. (2003), 'Perishing Paradigm: Race 1931-1999', *American Anthropologist*, Vol. 105, pp. 110-113.

Maxwell, A. (1999), *Colonial Photography and Exhibitions: Representations of the "Native" and the Making of European Identities*, Leicester University Press, London.

Mellinger, W.M. (1992), 'Postcards from the Edge of the Color Line: Images of African Americans in Popular Culture, 1893-1917', *Symbolic Interaction*, Vol. 15, pp. 413-433.

Miles, M. (1992), 'The Virgin's One Bare Breast', in N. Broude and M. D. Garrard (eds), *The Expanding Discourse: Feminism and Art History*, HarperCollins, New York, pp. 27-37.

Pauli, L. (2006), *Acting the Past: Photographs as Theatre*, National Gallery of Canada, London.

Ridpath, J.C. (1893), *The Great Races of Mankind: An Account of the Ethnic Origin, Primitive Estate, Early Migrations, Social Evolution, and Present Conditions and Promise of the Principal Families of Men*, Four Volumes, C.W. Slauson Pub. Co., Chicago.

Said, E.W. (2001), *Power, Politics, and Culture: Interviews with Edward W. Said*, edited by G. Viswanathan, Pantheon, New York.

Sontag, S. (1977), *On Photography*, Farrar, Straus and Giroux, New York.

Staff, F. (1966), *The Picture Postcard and its Origins*, Frederick A. Praeger, New York.

Stanton, W. (1960), *The Leopard's Spots: Scientific Attitudes toward Race in America 1815-59*, University of Chicago Press, Chicago.

Stocking, G.W. Jr. (1991), 'Maclay, Kubary, Malinowski: Archetypes from the Dreamtime of Anthropology', in G.W. Stocking Jr. (ed.), *Colonial Situations:*

Essays on the Contextualization of Ethnographic Knowledge, University of Wisconsin Press, Madison, pp. 9-74.

Tamplin, R. (1992), 'Noblemen and Noble Savages', in Mick Gidley (ed.), *Representing Others: White Views of Indigenous People*, University of Exeter Press, Exeter, pp. 60-83.

Van den Berghe, P.L. (1984), 'Introduction: Tourism and Re-Created Ethnicity', *Annals of Tourism Research*, Vol. 11, pp. 343-352.

Vanderwood, P.J. and Samponaro, F.N. (1988), *Border Fury: A Picture Postcard Record of Mexico's Revolution and U.S. War Preparedness, 1910-1917*, University of New Mexico Press, Albuquerque.

Whittaker, E. (1986), *The Mainland Haole: The White Experience in Hawaii*, Columbia University Press, New York.

Whittaker, E. (1999), 'Indigenous Tourism: Reclaiming Knowledge, Culture and Intellectual Property in Australia', in M. Robinson and P. Boniface (eds), *Tourism and Cultural Conflicts*, CAB International Publishing, Oxon, pp. 33-46.

Whittaker, E. (2000), 'A Century of Indigenous Images: the World According to the Tourist Postcard', in M. Robinson (ed.), *Expressions of Culture, Identity and Meaning in Tourism*, Northumbria University Press, Newcastle, pp. 425-437.

Chapter 7

From Images to Imaginaries: Tourism Advertisements and the Conjuring of Reality

Teresa E.P. Delfín

How we represent space and time in theory matters, because it affects how we and others interpret and then act with respect to the world.

David Harvey (1990: 205)

As tourism has risen to the top of the global economy – and especially with the industry having become the darling of the World Bank and the IMF[1] – tourism has worked its way into everyone's imaginations and realities. While there is much to say about the rise of the development of tourism, it is these latter terms – *imagination* and *reality* – that I will be exploring in this chapter. By focusing on advertisements portraying and promoting tourism, I will consider the ways that visual rhetoric influences the ways that we imagine and consequently construct tourism realities. Using an Althusserian framework, I argue that the ways we learn to imagine ourselves and others vis-à-vis tourism has a profound and lasting impact on the ways that tourism is experienced by both tourists and locals. Likewise, these imaginings also have consequences for the very landscapes that tourists occupy. While this is true for all tourism everywhere, I argue that the impact of tourism is felt most forcefully in developing[2] spaces. It is thus that the advertisements I consider portray Third World tourism specifically.

Although tourism has always been an imaginatively charged realm, perhaps the greatest contributor to tourism fantasies in recent decades has been advertisement. Not only has tourism been subject to immeasurable amounts of advertisement, but tourism itself has come to figure prominently as a backdrop for non-tourist related adverts as well. This has to do in part with the ubiquity of tourism, but perhaps more so with the imaginative qualities with which tourism has always been endowed. Because tourism inherently inspires fantasy, it is an ideal vehicle for the recruitment of potential consumers of goods that may or may not have to do with tourism. As I will argue, it is the imaginatively engaging and identity constituting process of recruitment, which is a precursor to consumption, that is much more important than consumption itself. In order for consumers to be recruited, they first

---

1   For a detailed discussion of the role of the IMF and the World Bank in tourism, see Enloe (pp. 120-121, 163-165, 183-189).

2   Throughout this text I use the terms "developing," "Third World," and "postcolonial" interchangeably.

have to pass through a phase of imaginings in which they reformulate the ways they imagine themselves to incorporate that which is being advertised, whether it be a thing, an experience, etc. While this might seem rather tedious and deliberate, this is a process that nearly every potential consumer goes through.

The process looks a lot like this: A catalogue arrives in the mail. Its recipient can either choose to discard it or to peruse it; this first act does not itself dictate whether the recruitment will be effective, but it is when the person chooses to consider the contents of the catalogue that potential recruitment begins. As he flips through the pages and beholds the images, his eyes linger over the things he finds interesting. He then begins to imagine his own life with the addition of a particular thing – let's say, for simplicity's sake, a coat. Does he need a coat? He could make that case. What about the color; are there choices? It comes in his favorite color, a color he knows he looks good in. What about the fabric? He hates acrylic, and pictures the way that his last acrylic coat fell apart prematurely and was unpleasant to the touch. Acrylic is definitely out. But this coat is wool, which he is fond of. He imagines wearing the coat – a navy blue wool pea coat with nondescript buttons (he hates shiny brass ones – they make him feel like he is a child playing at being a sailor). He imagines the way people respond to him in that coat, imagines traveling in it, imagines including it in his life. At this point, the man's recruitment as a consumer is nearly complete, subject only to the formalities of purchasing the coat. Then recruitment is accomplished. Not only has our subject been recruited to purchase the coat, but in the process of weighing the decision, he has also become "a wool pea coat kind of guy."

The imaginative process outlined above not only helps determine what kinds of goods a subject consumes, but this process is also a constitutive and creative factor in self-definition and differentiation. As Arjun Appadurai explains, "consumption is now the social practice through which persons are drawn into the work of fantasy. It is the daily practice through which nostalgia and fantasy are drawn together in a world of commodified objects" (1996: 82).

If anything, tourism only exaggerates the claims being made about consumption. The would-be tourist considers how the experience being offered correlates with what she does desire or could desire, and then takes this a step further to not only include herself in her image of the tourism site, but to also include the tourism experience in the image she has (or would like to have) of herself. As Judith Williamson explains: "Advertisements are selling us something else besides consumer goods: in providing us with a structure in which we, and those goods, are interchangeable, they are selling us ourselves" (1978: 13). Where tourism goes further than other commodities, experiences, or decision making that is influenced by advertisement is that tourism, even aside from its advertised dimensions, is a construct always-already replete with imaginative dimensions. Similar to the ways that advertisements ask us to behold potential commodities and imaginatively "try them on," tourism's requisite difference from our everyday lives likewise requires that we embark in an imaginative process – parallel to that instigated by advertisement – in order to determine whether to become a tourist, where, and to what extent. After all, tourism destinations are, if nothing else, places where people are encouraged to craft there own ideal temporary realities, often with few similarities to their everyday lives. A quick assessment of any tourist destination reveals that much of the appeal of tourism is in its difference from the everyday life being left behind. I don't entirely accept John Urry's pronouncement that what tourists ultimately pursue is an *inversion* of everyday life (1990). Instead, I agree with Nelson Graburn that, "tourists on holiday are seeking specific reversals of a few specific features of their workaday home life, things that they lack or that advertising has pointed they could better find elsewhere" (Smith and Brent 2001: 50).

While the things that tourists look for in their holidays may be specific, the landscapes in which these needs may be met are often very general, interchangeable, or even negligible. This is especially true in postcolonial tourism sites, where foreigners have long arrived and made their own meanings of spaces, in most cases ignoring history, geography, and topography. Ann McClintock argues for the recurring colonial trope of "empty land," or in a more gendered reading, "virgin land." She argues, "if the land is virgin, colonized people cannot claim territorial rights" (1995: 30). In tourism, I find it more useful to think of spaces not as empty and thus completely open to meaning making, but as palimpsests, with layers of meaning to negotiate and contribute to. In this model, it is the tourist who contributes the top layer of meaning and imaginings to the tourism palimpsest; while the other layers are still visible, they are easily ignored or acknowledged or hyper-valorized at will.

Tourist destinations are places where fantasies are superimposed on landscapes. While there is generally some official sanctioning of what tourism sites are meant to invoke ("paradise," "the past," etc), tourists arrive at their destinations with their own agendas, prepared to make *their* tourist sites conform to their fantasies. As in Althusser's notion of interpellation in which subjects are hailed – ultimately by and for ideology – in tourism, tourists are not only hailed (interpellated) as subjects, they also recruit their own subjects. But where Althusser's model accounts mostly for the interpellation of human subjects, in tourism, spaces themselves are also open to interpellation. Indeed, once a space is interpellated, it serves as a constant reminder to those who occupy it of the space's significance; this in turn contributes to a parallel process of mutual constitution between tourists and the spaces they occupy. In this process the tourist is responsible for actively crafting her destination into a place that corresponds with her fantasies. But, to achieve this, the tourist must *also* exhibit behaviors befitting her interpretation – or the interpretation by which she has been recruited. Likewise, the place itself will impose certain behaviors on tourists. At issue here is not the determination of whom constitutes what (or vice versa), but that landscapes and people are *both* heavily loaded with imagery, fantasy, and expectation, and as such *both* come to reflect each other indefinitely such that the tourist will begin to "act the part" effortlessly, and the tourism site will reflect that which is desired by the tourist: a virtual hall of mirrors.

I have seen this play out countless times while doing fieldwork on the Inca Trail to Machu Picchu (2001-2005). Tourists hit the trail with expectations of ruggedness and adventure, but would eventually be overcome by the material evidence of an Inca past that is impossible to ignore on the four-day hike. The Inca "presence" most commonly became translated as something mystical or spiritual, which invariably came to dominate many trekkers' experiences of the trail. Confirming the completion of the interpellation, on several occasions I heard trekkers be chastised by their friends for such things as listening to music through headphones while hiking – an act that was seen as normal on the first day of the hike, but which was called out for being not being meditative or spiritual enough by day three. Likewise, hikers repeatedly and increasingly equated the porters, who move quickly and quietly up and down the trail, with the Incas, and saw them as evidence of a sort of anachronistic continuity that fit squarely within the time-travel fantasy that walking to Machu Picchu regularly became.

A commonly deployed trope in Third World tourism is a temporal one in which space is undermined by time – specifically the past. In the tourist (and indeed colonial) fantasy of being able to experience spaces fixed in the past, the mutually constituting interpellation described above is everywhere present. If a tourist decides that what he wants out of his holiday is to experience life in some past, then to facilitate this, several elements have to come together: the landscape, above all, has to be able to pass as ancient, colonial, pre-historical, etc. Likewise, the inhabitants of these spaces are expected to look and behave "traditionally." If a tourist wants only to sample the past, he can treat his "time-travel" as he would a museum and *observe* the imagined past around him. This *museumization*,[3] has its roots in colonialism and is deeply connected to modernity's obsession with the temporal. David Harvey notes, "Since modernity is about the experience of progress through modernization, writings on that theme have tended to emphasize temporality, the process of *becoming*, rather than being in space and place" (1990: 205). Doreen Masssey argues that these temporal tendencies have not disappeared, but that in fact they have been incorporated into the feminist, postmodern, and postcolonial focus on the local. She writes, "'Place' in this formulation was necessarily an essentialist concept which held within it the temptation of relapsing into past traditions, of sinking back into (what was interpreted as) the comfort of Being instead of forging ahead with the (assumed progressive) project of Becoming" (1994: 119). This "sinking back," along with the reification of the local, are characteristic of a second option for the tourist bent on the idea of the past: that is "going native." This option, borrowed from anthropology, often stems from a rejection of the tourist's or anthropologist's own culture, and proceeds with the subject adopting, to the extent possible, the local culture.

---

3   Dean MacCannell explains, "the best indication of the final victory of modernity over the sociocultural arrangements is not the disappearance of the nonmodern world, but its artificial preservation and reconstruction in modern society" (1999: 8)

A third option is to "go colonial" and enjoy the privileges of the present and the past simultaneously. This last option almost always foregrounds inequalities between tourists and hosts, and in some cases this hierarchy is a selling point of such vacations. The notion of "going colonial," is also closely linked to Johannes Fabian's notion of allochronism in which, especially in anthropology, time has served as a "technology for othering" (1983). For Fabian, this othering takes place by imagining people as inhabiting what amounts to a permanent past relative to Westerners' present. Aihwa Ong brings this notion up to date, explaining: "The hegemonic Euroamerican notion of modernity – as spelled out in modernization theory and theories of development – located the non-West at the far end of an escalator rising toward the West, which is at the pinnacle of modernity in terms of capitalist development, secularization of culture, and democratic state formation" (1999: 31). Néstor García Canclini attempts to justify these temporal fantasies as follows: "In this epoch in which we doubt the benefits of modernity, temptations mount for a return to some past that we imagine to be more tolerable…the evocation of distant times reinstates in contemporary life archaisms that modernity has displaced" (1995). Underscoring this explanation, however, is a concern over the simultaneous displacement and desire for these archaisms. The struggle in place here brings to mind Renato Rosaldo's concept of Imperialist Nostalgia, in which "somebody deliberately alters a form of life, and then regrets that things have not remained the way they were prior to the intervention" (1989: 69-70). This is only one of the ethical dilemmas associated with "going colonial."

The idea of overlapping temporalities is a rhetorical staple of colonialism that has now been revamped by tourism. This temporal slippage allows for participation in vacation sites that take the most alluring or paradigmatic features of various temporalities and fuses them into one ideal time-travel imaginary with mass appeal. In what follows, I will discuss several advertisements that are consistent with this model in that all depict travel to developing regions situated as existing in the past; not surprisingly, two of these advertisements depict spaces in Latin America.

The first advertisement, for Marriott Hotels and AT&T reads: "Room 741. Wants to go back two centuries without giving up e-mail." "Room 741" refers to the guest who occupies the room, a statuesque blonde in unseasonable business attire who inspects hand-woven fabric. She stands towering over not only the all female weavers in indigenous dress who sit on the ground, but also over their presumptive boss, a jovial looking mestizo who must tip his head back to meet her eyes. While the ad's captions already goes a long and not-so-subtle way in establishing the advertisement's agenda, the smaller print caption below the image takes this rhetoric even further:

You've changed time zones, continents, even centuries. But that doesn't mean you
have to change the way you work. Our hotels in 34 countries have VOICEMAIL,
BUSINESS SERVICES like faxing, even easy to reach DATAPORTS. Call

your travel agent or 800-228-9290, or visit www.marriot.com. We believe:
When you're comfortable you can do anything.

While it is noteworthy that this advertisement is thick with contrasts (between the tall, foreign blonde and the diminutive people surrounding her; between technology and tradition; and between the present and the past), what is most noteworthy is the uncharacteristic way in which the portrayal of the time-spaces around which these contrasts are constructed seemingly contradicts the very logic that motivates the temporal fantasies that are being sold. Here I am referring to the tendency of looking to the past for stability, tradition, and authenticity, while the present has long been regarded as the site of chaos, rupture, malleability, and inauthenticity. Instead, this and the following ads insist on the reversal of such paradigms through the juxtaposition of the *past* as a site of escape and of metropolitan *modernity* as a site of duty or reality.

The ability to remain as connected as one would be on Wall Street while browsing colonial *puestos* gives the less adventurous traveler the ability to get out without getting over her head. This completely reconfigures the widely theorized notion that modernity is a site of inauthenticity constructed largely by consumerist escapism. It dutifully turns the logic *back* around to locate premodernity as a site for tourist escape or easy business. As such, it also renegotiates the relationship to time, making the *premodern* the time–space of dynamism while constructing modernity, the moment of seemingly permanent corporate habitation, as static. Indeed, one of the aims of such advertisements is to simultaneously cater to and attempt to *reverse* our reverence for the past, for out-of-the-way places, and for traditional people. Such ads show that not only are these categories no longer as pure as we thought, but that corporate culture may possess some of their attributes. Further, this ad challenges popular depictions of premodernity in which this temporal construct is feminized to symbolize originality, consistency, authenticity, and a location of homecoming and healing. But through the reversal of paradigms established within all these advertisements, premodernity is *also* suggested to be fickle and easily penetrated by the presence of foreign business, customs, money, and people – as an obliging time–space of frequent change and malleability. Needless to say, premodernity is defined against modernity, which, as inscribed by capital, wields the power to manipulate spaces and the ways they are perceived. Thus modernity and its subjects (or ambassadors), even when physically outnumbered, still come to dominate the surroundings against which they are positioned.

Both this and the following advertisement self-consciously recognize that there need be no authentic reconciliation between overlapping premodern and modern spaces; that in fact there is a ready clientele who revels in this simultaneity, ambiguity, and constructedness. "Post-tourist" is what Maxine Feifer calls those travelers who understand that authenticity is limited to them and thus revel in the constructions of authenticity that are part of the performativity of tourism (1985). The post-tourist recognizes the contradiction and performance inherent in the time-traveling imaginary and finds the act of performance of authenticity as interesting and engaging as authenticity itself would be. The post tourist is knowingly entertained by performance and simulacra, seeing the fruitlessness of searching for authenticity.

An AmericaWest advertisement (1998) – featuring a closely cropped photo of a stone carving – reads, "When you fly to Mexico, remember to set your watch back 1,000 years" and exaggerates the rhetoric of the previous ad. Both highlight the fraught mediation that is inherent to temporization. Both advertisements explicitly locate geographic spaces as situated in specific pasts while making apparent the fallacy (or fantasy) of such a claim through the mediation of modern artifacts. However, consistent with the logic of these conflated temporalities, computers and clocks call no bluffs. Instead, it is what these technologies represent that gives those who own and control them the privilege of time travel. In "Room 741," a specific past (two centuries back) is mediated and arguably negated by the unseen, yet referenced presence of a computer connected to the Internet through phone lines, cable, or satellite. In the AmericaWest ad, it is the wristwatch that comes to signify both modernity and modern human agency, as it is the conscious act of "set[ting] your watch back" that heralds the journey to a pre-modern past whose threshold is symbolized by the "ancient" carving. Because the wristwatch is symbolic both of the modern and of all time simultaneously, the assumption is that without possession of this artifact there is only access to one temporal realm – presumably the premodern time epitomized by the stone carving. Thus, as in the previous ad, it is clear that one of the perks of citizenship in this late-capitalist moment is the ability to move unhindered across space and time.

These artifacts of modernity, the internet and the wristwatch, by being foregrounded in advertisements whose claim is to package and sell the past, succeed only in negating the possibility that the pre-modern space – either real or imagined – be authentic in any capacity. Further, by privileging the unstable artifacts of postmodernity, the past*ized* space masquerading as a commodity is reduced to a sideshow. As concepts such as authenticity and tradition lose meaning through the empty portrayals to which they are submitted in these advertisements, notions such as stability and reliability gain momentum, as these terms are part of a high-tech currency. The result is the unmasking of the *real* commodities: mobility and connectedness. The final commodity that belongs in this series is of course communication itself, with its promises of the ability to inhabit two places and times simultaneously through the advent of technology.

The final advertisement privileges the same categories that have been present in the two previous images, while foregrounding the idea that dichotomy and distance are organic or authentic categories that must be mediated technologically. The caption for this AT&T ad (2000) reads: "How to get home fast from 1,000 years away." Here again, it is a blonde woman who is the focal point. Unlike the Marriott ad, however, the central figure here, seen riding a bike, appears to be college-age. The shorts, bright blue tank top, and backpack she wears – in contrast to the somber and conservative attire of the probable locals – give her away as a foreigner. The structure here is remarkably similar to that of the Marriott ad, though this is no surprise, as AT&T is also featured in that one. In both of these advertisements, those protagonists with access to the technologies advertised serve as avatars – no longer ambassadors – of the high-tech realm. In both of these ads the subjects bare the same markings as the technologies: they are white northerners who are beautiful, capable, adventurous, and inviting; and as stated earlier, both mark their backdrop as "foreign" by their own presence in it. We'll note as well that although (or perhaps because) the blonde women in both of the AT&T ads are the misfits in the spaces they occupy, they each nonetheless become the focal points. But while the presumed locals gaze at the blonde on the bike, *she* gazes at *us*. It is this direct engagement that serves to hail us, to make clear that the message in the caption is not only meant for her – the blonde – but also for us, her symbolic equivalents. Judith Williamson explains this process as follows:

> [Advertisement] exists in and out of other media, and speaks to us in a language we can recognize but a voice we can never identify. This is because advertising has no 'subject.' Obviously people invent and produce adverts, but apart from the fact that they are unknown and faceless, the ad in any case does not claim to speak from them, it is not their speech. Thus there is a space, a gap left where the speaker should be; and one of the peculiar features of advertising is that we are drawn in to fill that gap, so that we become both listener and speaker, subject and object (1978: 13-14).

The directions in which these gazes travel – both in and out of the advertisements – is further indication of the purpose of the technologies being advertised in such foreign spaces. As these advertisements suggest, these technologies exist abroad not so much so that locals can use them, but so that we can inhabit as many worlds as we want while away from home. Rhetorician Cynthia Selfe comes to a similar conclusion in an analysis of advertisements for exports of technology to developing countries: these ads serve to reinforce an already well-established script that originates in our history of "experiencing the world as missionaries, as colonists, as tourists, as representatives of multinational companies." However reconfigured, the result is the same: "Americans are the smart ones who use technological expertise to connect the world's peoples, to supply them with technology and train them to use it" (1999: 295). In Selfe's analysis, these technologies are made available to non-Westerners, but at the expense of American dependency.

As the advertisements I have discussed exemplify, dreams of tourism rarely ever invoke the commercial cliché of a solitary beach chair on a private beach where you could be anywhere or nowhere.[4] While the space's specificities may

---

4   The symbol of the beach chair is so ubiquitous that it is also the symbol for the *Journal of Tourism and Cultural Change*. Here, though, this symbol undergoes an ironic makeover as the beach chair is perpetually located "out of context" in dreary, wintry scenes, industrial zones, and other nontraditional venues.

be thoroughly reconceptualized by tourists, they are never empty or socially isolated. Instead, tourist imaginaries, as mentioned earlier, consist (at least) of the following categories: self, space, and others.[5] What's more advertisements such as the ones above not only recruit tourist subjects (their desired audience), but these subjects in turn embark on a complex chain of recruitments that all work towards the fulfillment of tourist fantasies. Let me be clear, however, that in using terms such as *fantasy* or *imaginary*, I am not referring to things without material manifestations or consequences. Instead, following Althusser I argue that, especially because of the already fantastic nature of tourism, tourist imaginaries do indeed have material consequences for the spaces tourists occupy and on the people with whom they may or may not come into contact but with whom there is nonetheless a connection based on the economic structures of tourism. Althusser explains this as follows:

> [A]ll ideology represents in its necessarily imaginary distortion not the existing relations of production (and the other relations that derive from them), but above all the (imaginary) relationship of individuals to the relations of production and the relations that derive from them. What is represented in ideology is therefore not the system of the real relations which govern the existence of individuals, but the imaginary relation of the individuals to the real relations in which they live (1971: 155).

In tourism, not only does the realm of the imaginary offer insights into real relations of production – it also helps create and reinforce these relations. If the work of advertisements such as those above is (in part) to recruit tourists to a temporally structured way of thinking about self, place, and other, then, as Althusser explains, there must be material consequences for such ways of imagining – and there are.

If a tourist wishes for her holiday to mimic the past then what is required of those who make this fantasy possible is one of two things: either the place and the people acting as hosts must live much as they did centuries back, or there must be technologies and scripts in place to convince the tourist that she is occupying her desired time and place.[6] In either case, hosts are being asked, for the sake of tourists' fantasies, to be either premodern or postmodern, but nothing in between. García Canclini explains, "it is possible to think that to be modern no longer makes sense at this time in that the philosophies of postmodernity disqualify the cultural movements that promise utopias and foster progress" (1995: 1). While such a stance may be acceptable in developed, late-capitalist areas that claim to have achieved the goals of modernity, and modernization, this is far from the case in most postcolonial regions. As Renato Rosaldo explains in his preface to García

---

5   Tourist imaginaries may of course also include categories such as language, adventure, food, sex, etc.

6   For a more detailed discussion of this phenomenon see Dean MacCannell's essay, "Staged Authenticity" (1999).

Canclini's book, "In Latin America, modernization and development remain vital issues that are named as such in the discussions that reflect and create national self understanding" (xiii).

In response to these temporal fantasies, I can't help but imagine Third Worlds aching towards modernization while having to maintain, for the sake of tourists' dollars and euros, that they are not modern at all, but certainly premodern enough to attract tourists, and postmodern enough to keep them safe and comfortable. Postcolonial tourist destinations do not themselves decide what's best for the lives, businesses, prosperity, and integrity of its citizens. Instead, these spaces, like all subjects (in the Althusserian sense), are hailed into being by a multiplicity of forces ranging from those as innocuous as individual tourists, to those as pervasive as the international financiers who benefit from such arrangements. In a poignant analysis of the political and gendered stakes of tourism, Cynthia Enloe notes: "Tourism is being touted as an alternative to the one-commodity dependency inherited from colonial rule. Foreign sun-seekers replace bananas. Hiltons replace sugar mills" (1990: 31). It is, of course, to the benefit of tourists not to inquire what happens backstage (MacCannell, 1999: 91-107) – whether the safe drinking water available to guests is likewise made available to locals is but a minor example. Likewise, tourists are discouraged from probing too deeply the source of their ever-lasting dollar abroad – especially in places where the national currency is often worth less than one thousandth of their own imported bills. Enloe is particularly attentive to the role that Third World tourism plays in the global economy:

> Tourism entails a more politically potent kind of intimacy. For a tourist isn't expected to be very adventurous or daring, to learn a foreign language or adapt to local custom. Making sense of the strange local currency is about all that is demanded. Perhaps it is for this reason that international technocrats express such satisfaction when a government announces that it plans to promote tourism as one of its major industries. For such a policy implies the willingness to meet the expectations of those foreigners who want political stability, safety and congeniality when they travel. A government which decides to rely on money from tourism for its development is a government which has decided to be internationally compliant enough that even a woman traveling on her own will be made to feel at home there (1990: 31).

The compliance to which Enloe refers marks the completion of the process of interpellation. A scenario such as the one Enloe describes indicates that spaces demarcated for tourism have made themselves open to interpellation by international tourists and corporations. At this point, sites slated for tourism development begin a long process of being inscribed with new meaning. This is a process that includes participation by an unlimited number of players to innumerable ends. While there is always the possibility that the imaginings attached to tourism spaces are perverse or destructive – and examples of the latter abound, I am nonetheless optimistic that a vital side of the process of tourism interpellation is not only imaginative, but also potentially generative and recuperative.

Works Cited

Althusser, L. (1971), *Lenin and Philosophy and Other Essays*, Monthly Review Press, New York.

America West Airlines (1998), Poster Advertisement.

Appadurai, A. (1996), *Modernity at Large: Cultural Dimensions of Globalization*, University of Minnesota Press, Minneapolis.

AT&T (2000), *Student Travels*, Advertisement.

Enloe, C. (1990), *Bananas, Beaches, and Bases: Making Feminist Sense of International Politics*, University of California Press, Berkeley.

Fabian, J. (1983), *Time and the Other*, Columbia University Press, New York.

Feifer, M. (1985), *Going Places*, Macmillan, London.

García Canclini, N. (1995), *Hybrid Cultures: Strategies for Entering and Leaving Modernity*, University of Minnesota Press, Minneapolis.

Harvey, D. (1990), *The Condition of Postmodernity: An Enquiry into the Origins of Cultural Change*, Blackwell, Cambridge, MA.

MacCannell, D. (1999), *The Tourist: A New Theory of the Leisure Class*, University of California Press, Berkeley.

Marriott and AT&T (Oct. 6, 1998), *Newsweek*, Advertisement.

Massey, D. (1994), *Space, Place and Gender*, University of Minnesota Press, Minneapolis.

McClintock, A. (1995), *Imperial Leather: Race, Gender and Sexuality in the Colonial Contest*, Routledge, New York.

Ong, A. (1999), *Flexible Citizenship: The Cultural Logics of Transnationality*, Duke University Press, Durham.

Rosaldo, R. (1989), *Culture and Truth: The Remaking of Social Analysis*, Beacon, Boston.

Selfe, C. and Hawisher, G. (eds) (1999), *Passions, Pedagogies, and 21st Century Technologies*, Utah State UP, Logan, Utah.

Smith V. and Brent M. (2001), *Hosts and Guests Revisited: Tourism Issues of the 21st Century*, Elmsford, Cognizant Communication Corporation, New York.

Urry, J. (1990), *The Tourist Gaze: Leisure and Travel in Contemporary Societies*, Sage, London.

Williamson, J. (1978), *Decoding Advertisement: Ideology and Meaning in Advertising*, Marion Boyars, London.

Chapter 8

The Camera as Global Vampire: The Distorted Mirror of Photography in Remote Indonesia and Elsewhere

Janet Hoskins

The concern with securing photographs without the knowledge of the subject being photographed is perhaps an appropriate metaphor for the ambition of regimes of colonial representation: to see without being seen. As Timothy Mitchell suggested, in the context of European colonialism in the Middle East, 'the photographer, invisible beneath his black cloth as he eyed the world through his camera's gaze...typified the kind of presence desired by the European whether as tourist, writer or indeed colonial power'. James Ryan, *Picturing Empire* (1997: 144)

This chapter probes the old colonial stereotype of the native who 'fears that the camera will steal his soul' and tries to provide both a historical context for these accounts and a theory of how they may have come into being. It also presents new ethnographic materials which – in combination with other recent studies in a postcolonial context – suggests an alternative phantasm of global vampirism, which has been abundantly documented in the last decade.

Many observers have been skeptical of stories of credulous savages who were unable to distinguish the technology of photography from their own fantasies of demonic spirits. But they have also acknowledged that such stories are widespread, that they are found from the Middle East and Africa to China, from Brazil to Indonesia, and that they must therefore correspond to some sort of discursive formation in colonial thought and perhaps even to recurrent behaviors on the part of those being photographed. While these behaviors may have been misinterpreted, and while 'animistic fantasies' may characterize the Europeans as much as 'the natives', it is important to try to formulate what goes into them, and also ways in which acts of visual appropriation can be turned around.

Accounts of Predatory Cameras

Colonial explorers and photographers delighted in telling stories about how they terrorized natives with a camera. The wife of photographer Louis Agassiz reports on native reactions to being photographed in Brazil in 1868: 'There is a prevalent superstition among the Indians and Negroes that a portrait absorbs into itself something of the vitality of the sitter, and that any one is liable to die shortly after his picture is taken. This notion is so deeply rooted that it has been no easy matter to overcome it' (Banta Hinsely 1986: 58). Some of this may have due to the length of time that they had to stand still, the intimidating looking apparatus, and the fact that they had no idea what he would do with the images.

In 1878-80, a British photographer on the East Central African Expedition claimed that 'by leaving a camera standing alone he kept a whole village totally deserted for a day' (Thompson, cited in Ryan 1997: 143). Another traveling in southern China asserted that his camera

> was held to be a dark mysterious instrument, which, combined with my naturally, or my supernaturally intensified, eyesight gave me power to see through rocks and mountains, to pierce the very souls of natives, and to produce miraculous pictures by some black art, which at the same time bereft the individual depicted of so much of the principle of life as to render his death a certainty within a very short period of years (Thompson 1874: 3).

Contemporary commentators on these stories have, understandably, often voiced words of caution about taking these stories literally. Alison Griffiths (2002: 105) notes 'Repeated reminders of the challenges or risks involved in taking photographs of native peoples should be read with a certain degree of skepticism, though, since inflated egos and exaggerated accounts of personal challenges were very often the result of self-aggrandizement rather than documents of fact'. And James Ryan expands on this point when he says (1997: 143) 'general suspicions of the camera and its operator were not perhaps entirely unfounded, since those sitters whom photographers did manage to capture had neither knowledge nor control over the uses and meanings of their likenesses. Thus to a large extent explorer's accounts of superstitious fears of the camera were themselves misreadings of cultural difference and the very real threat that their presence and technology posed'.

Imputations of native superstitions could have served to excuse bad behavior on the part of the picture takers, who may have failed to work to win the trust of local people, or to compensate them for intruding into their homes, enduring physical discomfort, and giving generously of their own time.

But such stories are not confined to the now vanished world of colonial empire. In 2000, when I returned to the eastern Indonesian island of Sumba (an island I have visited for nine different fieldwork studies from 1979 to the present), I encountered a new version of this old colonial cliché: 'Foreigners with metal boxes' (*dawa mbella*) were feared in remote villages because they were believed to use the hose-like aperture of their zoom lenses to extract blood from children and use it to power electronic devices in their own homelands. The blood was extracted by hanging the children upside down, opening a hole in their heads, and draining the blood into the metal box. I collected versions of this story from forty different people, from children to old people, illiterate villages to educated – and

Figure 8.1   Children watching video tapes in a Sumbanese village in 1988
Source: Janet Hoskins, with the permission of Nelden Djakababa

skeptical – town dwellers. I recognized the theme so often announced in colonial narratives, but was immediately and painfully aware that I was rediscovering it as an anthropologist who spoke the local language, had spent over three years in the field, and did not consider herself a peddler of imperial fantasies about native credulity. The particular form that this story took is revealing: Sumba was a very isolated island which was never targeted for economic exploitation during the colonial period, and only came under Dutch control in 1915. The first archival photographs we have of the region are from 1949, when it was visited by the Swiss ethnologist Alfred Buhler. He did not stay to study the languages and cultures, but did travel to remote areas with a large box-like camera and tripod. In the 1970s and 1980s, when I did the bulk of my own field research, it was still not possible to develop film on the island, and village people did not own cameras of their own. All Indonesian citizens had been photographed by government teams that issued identity cards, and many had a few treasured snapshots given to them by visitors, missionaries or government officers. In the 1990s, paved roads were built into regions like Kodi and a once monthly boat began to visit the island as well as tiny planes flying from Bali twice a week. Sumba was written up in *The Lonely Planet Guide* and the *Guide Routard* as one of 'the *in* places to visit' for backpacker tourists, and thousands of young people carrying cheap clothes and expensive cameras hiked into Sumbanese villages. This is when the stories of 'foreigners with metal boxes', said to roam in great numbers in the months of July and August, became prevalent.

It is possible to imagine how the story of the blood sucking camera could come into being: Even today, the click-clack sound of the mirrors of a single lens reflex camera falling into place is interpreted by many as an aspiration, a 'sucking sound'. Children are the most enthusiastic subjects for photographers in remote areas, since their thirst for novelty – and handouts like pencils, candy and coins – outweighs the suspicious reserve of many of their elders. It seems probable that some early photographers – perhaps even Buhler himself – tried to explain what they were doing by allowing children to look into the box and see the inverted image inside – a small dark room in which children were visibly hanging upside down. For Sumbanese, animists and Christians in a nation which is 85% Islamic, the image of children hanging upside down would suggest the required Islamic mode of sacrifice and butchering, where the animal's throat is first cut and it is then hung upside down to drain the blood. And in the eighteenth and nineteenth centuries many Sumbanese children were kidnapped and taken away from the island as slaves by Muslim sailors (Hoskins 1993), so the idea of predators who go after children has a strong historical resonance (See Figure 8.1 for a scene of children watching video cameras in 1988).

I have tried to survey the literature of reports about mystical or dangerous powers attributed to cameras in remote locations, and have at this point three initial findings:

1. The first is that when reports come from responsible ethnographers or other long term residents who learned the local languages, an idiom of bodily extraction (the stealing of a vital fluid – like blood, in many parts of Southeast Asia, or fat, in many parts of Latin America) predominates over ideas of 'soul stealing', which may be Christian-influenced interpretation of more visceral local imagery.[1]

2. The second is that the camera is not alone in provoking these reactions, although it is certainly the primary image of predation associated with tourism. Other forms of technology which can inspire both terror and admiration include the firemen's hoses in many African cities, which have been associated with an elaborate tradition of urban vampirism (White 2000), the hypodermic needle reported to provide stores of blood or fat for mysterious uses in Andean villages (Taussig 1987, Wachtel 1994) and colonial Uganda (White 2000), and the game ranger's rifle in Kenya and Tanzania (White 2000) – a use also paralled by the ritual use of rifles on Sumba during the period of colonial conquest (Hoskins 1986, 1993, 1996). The very enchantment with technology that has provoked these stories has also produced a form of resistance which I call 'visuality in reverse' – a

---

1   The imagery itself, of course, is highly variable and often related to inversion of practices ascribed to the local population – in Southeast Asia, this produces stories of 'white headhunters', while in China it may be 'foreign devils', and in Australia, it may be nocturnal witches who feed on the liver.

privileged use of cameras to valorize local perceptions and traditions in ways which differ from tourist consumption practices.

3. The third is that emphasis on natives as the objects of cameras has obscured the great importance of photography in domestic tourism throughout Asia, and the many ways in which photography is creatively reinvented in the non-western world.

Global Vampirism versus the Soul-Stealing Camera

Stories about predatory cameras collected by anthropologists rarely provide direct evidence for the idea of soul stealing. Most present a more physiological image of extraction. Sir Baldwin Spencer and F.J. Gillen were told by an old man in central Australia that 'our object in taking photographs was to extract the heart and liver of the blackfellows' (Gillen and Spencer 1968: 261). When David MacDougall tried to photograph Turkana women in northern Kenya in 1973 he was told that they were afraid the camera 'might make us weak' or 'make our blood thin' (MacDougall 1992: 104). Blood is required for virtually every act of empowerment in societies which practice sacrifice (such as the Turkana of Africa, or the Sumbanese in Eastern Indonesia) so that it is assumed to be vital fluid needed to assure the proper functioning of modern devices as well. The specific form that each set of stories takes builds local traditions (ideas of white headhunters, construction sacrifice, vampires, etc.) connected to earlier experiences of predation and exploitation.

The experience of watching indigenous people regard a camera with suspicion, turn away, or refuse to pose is extremely common. But many people in modern industrial countries also experience the intrusiveness of a stranger's camera, and may resist the voyeurism of the paparazzi or the callous journalist. It therefore seems possible that the imposition of a theory of 'soul stealing' is an almost purely Western fantasy. There have many anthropological critiques of abuses of the idea of the 'soul' in interpretations, usually presented by Protestant writers trying to make sense of indigenous practices. Rodney Needham's 'Skulls and Causality' – one of the best known – argues that there is no need to posit a belief in 'soul stuff' to understanding Southeast Asian headhunting. In a similar vein, Roger Keesing has written against the 'celestialization' of indigenous ideologies which sometimes makes them seem exotic and romantic rather than pragmatic and materially based (1984). Webb Keane argues that the very materiality of Indonesian exchange systems was hard for Protestant missionaries to accept (1994, 1997), since they kept looking for a transcendent 'soul'. While many peoples do have complex ideas of multiple souls and of soul loss, the very consistence of the 'soul stealing camera story' makes it suspicious.

Many other researchers have noted that photographing sacred objects is seen as risky. In Eastern Indonesia, Howell (pers. comm. 2003) notes that the Lio people of Flores told her that this would be dangerous both for the camera (either the pictures would not come out or the camera would be destroyed) and also for the sacred object (whose potency would somehow be reduced by representing it outside its proper context).

Smedal (1989) noted that the Bangka people of western Indonesia believed that electricity could be made by tapping the energy of newborn babies placed in acid. The children's natural vitality could be drained from them to make a new 'fuel' which would be used to power artificial light and generators used by foreign intruders. While these beliefs do not make specific reference to cameras, the idea that the children are bathed in dangerous chemicals to extract their vitality suggests the mysteries of the dark room, in which images of children appear in an acid bath as light sensitive paper absorbs their shadowy images.

MacDougall describes this set of stories as 'threatening', and groups them with the anthropometric photographs used to identify races and physical types, which are cases of 'photographic representation as violence' (MacDougall 1992: 107). But he also notes that there are also many stories of photography – especially in the third world – which are 'genially additive'. These stories of people dressing up for photographs, often in borrowed clothes, and using stage props (books, furniture, motorcycles) which may make them appear wealthier than they are. His research in the hill station of Mussoorie in the late 1980s and Christopher Pinney's work in Madhya Pradesh (2002) have provided extensive documentation of Indian appropriations and even re-inventions of photography, which are paralleled – with some important differences – in the Indonesian islands.

Visuality in Reverse: Resisting with a Camera

The potential of photography to both document reality and create fantasy images suggests that it is capable of serving a variety of ends, and the dominating gaze of colonialism can also be mocked and even appropriated by those who were once dominated. Dutch and British imperial masters helped to bolster and perhaps even at times create a 'native aristocracy' whose costumes were carefully constructed to fit roles assigned to them by Europeans (see Cohn 1990 on 'how the English dressed the Indians'), but at the same time they unwittingly participated in the process of creating new forms of glamour and enchantment. As Pinney argues:

> alongside the enforced visibility induced by photography lay a parallel process of making the Orient unknowable which, far from reinforcing the hegemony of colonial discourse, helped create an area of secrecy and power in which the "Other" could assert its own autonomy, and which was ultimately necessary for the establishment of the nation state (Pinney 1990: 284).

An Indonesian example of 'resistance through photography' could be the story of Yoseph Malo, the Raja of Rara from 1925-1960 who rose to this position from a life which included being sold as a slave as a young boy, taking heads himself to avenge his beheaded father, and being the first Catholic convert on the island

Figure 8.2   Three images of Yoseph Malo: the photo taken in 1957
Source: See Figure 8.1.

of Sumba. Malo was photographed on one day in his entire life, in 1957 (Figure 8.2), when his son graduated from high school and a Catholic priest brought a camera to the occasion. The three photos taken of him that day were then copied into a painting (Figure 8.3) for his belated wedding to his eighth wife, and finally for his grave and ultimately for the cover of the book (Figure 8.4) his son wrote about him. Each image was different (just him and his wife, just him, just him rejuvenated a bit and with his eye open, etc.) The newly-opened eye presents us paradoxically with a new gaze, from a man born in the nineteenth century and sold as a slave, we have a glance outwards from the colonial raja who looks out at his descendants in a cosmopolitan, globalized world.

Another example could be the photograph of the leaders of a resistance movement, led by the Kodi headhunter Wona Kaka, who were exiled to Java in 1913. Dressed in Dutch white shirts, their heads bound in red cloth and their shoulder draped with traditional textiles, these rebels had their images documented when they got off the white ship that took them to the eastern port of Surabaya. Forty years later, a descendant of theirs from the headhunting village of Bongu, Lota Mahemba, saw the photograph framed in Hotel Surabaya in Jakarta. He asked his employer, Raja Horo of Kodi, to negotiate with the hotel owner to purchase it and bring it back to Sumba. Wona Kaka's body had been lost, and his descendants had no idea where he was buried, so there were plans to use the photograph (along with his betel bag, head cloth and bush knife) in a ritual to call

Figure 8.3   The 1960 painting inspired by the 1957 photograph
Source: See Figure 8.1.

back his soul to the island and allow him to rejoin the community of ancestors in his home village. Since Wona Kaka was canonized as a hero of the national resistance in the 1950s, the Indonesian state offered to contribute to the cost of this ritual, merging ancestor worship with nationalist propaganda at the end of the Suharto era (Hoskins 1989).

Technologies which promise to copy the material world, or at least to present two dimensional images of it which we can recognize, are fascinating when they are introduced – photography, for instance, or film and television – but they can also be suspect. Benjamin argued:

> The authenticity of a thing is the essence of all that is transmissible from its beginning, ranging from its substantive duration to its testimony to the history which it has experienced. Since the historical testimony rests on the authenticity, the former, too, is jeopardized by reproduction when substantive duration ceases to matter. And what is really jeopardized when the historical testimony is affected is the authority of the object (Benjamin 221).

Benjamin's comments alert us to the unreliability of photographic images in relation to a more complex and layered real, and he notes that this is increased by the

RAJA YOSEPH MALO
OF SUMBA

Cornelius Malo Djakababa
Introduction by : DR. Janet A. Hoskins

Figure 8.4   The 2001 book cover, with his eye opened
Source: See Figure 8.1.

technological improvements in precision and fidelity of photographs, which have also become more subject to manipulation and 'lying'. As Spyer says: 'Photographs thus highlight their own incapacity to represent – a problem which, following Benjamin, must remain at the core of our technologies of representation' (2002: 192).

Gell has argued that anthropologists of art must move away from the study of indigenous aesthetics (which he calls the 'theology of a religion of high art') in order to study art as a technical process:

> The power of art objects stems from the technical processes they objectively embody: the technology of enchantment is founded on the enchantment of technology. The enchantment of technology is the power that technical processes have of casting a spell over us so that we see the real world in an enchanted form. Art, as a separate form of technical activity, only carries further, through a kind of involution, the enchantment which is immanent in all kinds of technical activity (1999: 163-164).

The inspiration that I drew from this section went in almost exactly the opposite direction from the one intended: Gell makes an argument that not only Trobriand canoe-prows but also modernist paintings are created in order to 'cast a spell over us' and allow us to 'see the real world in an enchanted form'. So, he specifically produces a theory of the social production and circulation of art objects which applies to both Western and non-Western art. In looking at the problem of reports about soul-stealing cameras, Gell's theory provided me with the insight that there is nothing irrational or even particularly 'primitive' in seeing the camera as a technology of enchantment – all forms of visual representation share this trait. What was distinctive about the situation of the tourist in a remote location was that the tourist had access to this technology while the native photographed did not.

The social life of technology affects the kinds of 'enchantment' which we may project onto it (in the sense of the old Hollywood cliché of 'movie magic' – which usually refers simply to bringing viewers closer to a glamorous other world) as well as the kind of 'spell' we believe it places on us (which in the case of Sumbanese villagers was predatory and extractive).

These new ways of using media challenge the rather more pessimistic predictions of an earlier generation of commentators. Edmund Snow Carpenter, who worked closely with Marshall McLuhan, believed that mass media would 'swallow up' indigenous people and take away their voices completely. He assumed that they would not be able to resist the power of these media, which would completely drain them of their vitality and cultural integrity. In the sixties and seventies, he experimented by giving villagers in New Guinea still and video cameras, and was disappointed in the results:

> Since around 1960, I've put cameras in a variety of hands. The results generally tell more about the medium employed than about the cultural background of the author or cameraman. In each case I had hoped the informant would present his own culture in a fresh way and perhaps even use the medium itself in a new way. I was wrong. What I saw was literacy and film. These media swallow culture. The old culture was there all right, but no more than residue at the bottom of a barrel. I think it requires enormous sophistication - media sophistication - before anyone can use print or film to preserve and present one's cultural heritage, even one's cultural present (1972: 185).

Thirty years later, this pessimism no longer seems as justified. While Carpenter thought that media themselves were 'invisible environments which surround and destroy old environments' (1972: 185), we now see many imaginative new uses of cameras in the hands of former colonial subjects. In many ways, Carpenter's predictions sound dangerously close to the fantasy of the 'little metal box' as an instrument of predatory extraction.

Photographs, which have become the pre-eminent material icons of the past for most of us, are also overwhelmingly identified with modernity. Deborah Poole (1997) has highlighted the ways in which an Andean 'visual economy' has organized a transnational flow of images and discourses of images between Peru and Europe, allowing the commodity status and capacity of photographs to accrue value as well as convey meaning in unequal exchanges. She joins with people like Ann Stoler (2002) and Rudolf Mracek (2002) in arguing that the colonies were often a testing ground for new technologies of visual control and expression, not merely a reflection of European developments but often a pioneer in producing them. From 1860 onwards, the colonies were depicted in *'cartes de visites'* – 'the sentimental green-backs of civilization' (Poole 1999: 109), as a sort of currency of global cosmopolitanism deeply invested in images of the primitive.

For the moment, the intense interest that former Asian colonial peoples, Latin Americans, and even Native Americans (Broder 1990) have shown in inverting the circumstances of the 'imperial spectacle' and in playing at dressing as their ancestors, the noble families who once ruled over them, and even their own colonial masters, should reveal something interesting about the camera as a technology of enchantment which also engenders an enchantment with technology.

Issues of visual appropriation and the ethics of visual documentation lie at the heart of the supposedly 'animistic' fantasy of the soul stealing camera. Why should third world peoples lend themselves without compensation to the tourist's apparently inexhaustible urge to 'capture' them in photographs? Why should certain groups always be the 'objects' of the camera's gaze while others are the 'authors' of exotic photo albums which show what daring, adventurous people they are? What rights do people have when the visual spectacle of their daily lives is a commodity which is marketed to eco-tourists and travelers throughout the world?

These ethical issues are sometimes, of course, addressed by anthropologists and even by tourist agencies, and they are hard to resolve in situations of restricted mechanical visuality. The gift of photographs as souvenirs, video tapes of film footage shot and other forms of 'visual sharing' can help, but I am sure many fieldworkers have had my experience of an almost unquenchable thirst for such images. After handing a full set of photographs of an event to the host, for instance, I am often deluged by requests for copies from each person who was a participant. After bringing video copies of our 1988 film to the principal persons featured in it, I was asked by the daughter of one to send her copies of all the unused footage as well. Gifts of tape recorders and still cameras prompt new requests for camcorders and VCRs. The addiction that Susan Sontag described as 'our obsessive compulsion to document our own lives in photographs' (1977) has already spread across the world. As time passes, these instruments will gradually become more widespread in the communities we study, and we can then come to recycle images they themselves have captured.

The camera can be 'enchanted' in two quite different ways, as I have tried to suggest in this chapter: (1) It can appear predatory and extractive, a dangerous tool for taking something of vital importance away from the subjects of its gaze, or (2) it can be a vehicle for re-imagining reality, creating playful possible worlds or even dream fantasies which offer substitute forms of gratification. I suggest that when there is little or no reciprocity between the 'picture taker' and the 'pictured', the first form of enchantment will predominate. The camera will be a mystified object of power that seems purely exploitative. When there is this reciprocity and sharing however (even in the commercial context of a photo studio or a 'camera for hire') then the camera opens up new possibilities for the creation of dream worlds and alternative visualities. This is because there is a chance for another intentionality to guide the camera, and for a co-creation of images rather than a simple posing for the surveillance of the colonial power. Once some control is put back into the hands of local photographers, however, the camera becomes 'enchanted' in a new way – as a tool to transform local vision and to create new dream worlds and play games with traditional visual systems. This is the way that cameras are being used now, as 'Western commodities become the objects of alien words' (Sahlins 1988: 384).

The world of modern industrial consumerism appears to us now as 'disenchanted', using Weber's term in the widest possible sense, to capture the idea that even notions such as the Hollywood idea of 'movie magic' are so thoroughly commercialized that they are rarely convincing. We know we are being manipulated and controlled to feel certain things in front of certain images, and we resist these manipulations with the 'weapons of the weak' most readily available to the reluctant audience – sarcasm, boredom, an unwillingness to let our imaginations be moved in puerile directions. But there is a somewhat different attitude in much of the third world, where exposure to new media is still relatively recent, and where finally gaining access to them in a way which allows the visual imagination an opportunity to exercise itself in new ways is a liberating experience, and one capable of many new enchantments. The new craze for photography in the third world stems from a global political economy in which mechanical visuality is restricted to certain peoples and certain institutions, and these lines of access are marked by differences of race and culture as well as class.

This tremendous new interest has produced new ways of using photographs and new ways of framing images. In many ways, these new usages follow Marshall Sahlins' dictum that the first thing that indigenous peoples do when confronted with commercial technology is not to be become more like 'us', but to use it to 'become more like themselves'. 'They turn foreign goods to the service of domestic ideas, to the objectification of their own goods and notions of the good life' (1994: 388). Indonesian photo studios specialize in invented images where people 'dress up like their ancestors' in front of digital cameras. Indonesian film makers like Garin Nugroho use Western media icons like Elvis and Madonna as alter-egos for local characters, and as devices to comment on the estrangement that people on remote islands feel towards Javanese 'internal colonialism' (Hoskins 1995). An American Filipino finds archival images from the World's Fair in St. Louis to tell another story of imperial domination and fragmented identity framed as a search for his own father (*Bontoc Eulogy*, 1997).

Photography and Domestic Tourism: Natives in 'White Face'

Costume photography in outdoor sites, like Gun Hill in Mussoorie, India (MacDougall 1992) or the Botanical Gardens in Bogor, Java is explicitly linked to a kind of tourism in which domestic tourists dress up like European colonizers (often in wedding dresses) or kings and queens (in ethnically coded traditional dress, *pakaian adat*).² In most cases in Indonesia, people are dressing up as an improved version of their own ancestors, with all the pomp and pageantry that was available only to hereditary nobles during the colonial period. So the most popular (and also the least expensive) form of costume used in the 'Tati Traditional Costume Photo Studio' (*foto studio adat*) in Denapasar, Bali, is a version of costumes worn by Balinese royalty. Other new costumes available include traditional Javanese, Chinese and European formal dress, 'classic' Dutch colonial, American cowboys, and now the Hindi film star costume (Dwyer 2002: 1). In this way, people can experiment with 'ethnic cross-dressing' – choosing to look like a Javanese prince and princess, for instance, even if they are really members of another regional group. Domestic tourists can also pose with both 'traditional' accessories (keris daggers, violins, fans) and 'modern' ones such as motorcycle, a guitar, glasses or a fake gun.³

A visit to a Traditional Costume Photo Studio is moderately priced (Rp. 25,000 or about $12), and includes the opportunity to pose in front of a number of backdrops, from a traditional Balinese split gate to a romantic sunset to a colonial scene complete with antique typewriter and telephone. Most customers are young, unmarried couples who want a romantic portrait to immortalize their love while it lasts. But there are also families commemorating anniversaries, bosses 'dressing like a king' to celebrate a promotion, and people who have just bought new homes (Dwyer 2002). A more complete (and more expensive) experience of colonial nostalgia for post-colonial 'natives' is available at the Bogor Botanical Gardens in Java, which were the pride of the Netherlands Indies and are still the sixth largest botanical gardens in the world. The Botanical gardens are the home of the

---

2   Many Javanese marriage customs were abandoned in the 1950s, when young nationalists and communists sought to replace the 'feudalism' of traditional social hierarchies with modern notions of freedom and equality. In the 1970s, under the autocratic New Order, supposedly 'authentic' and 'complete' Javanese wedding ceremonies were revived in the name of national pride, but since noble families no longer had the power to enforce their exclusive rights to such ceremonies, the court styles of nuptial regalia became available for rent to members of the new middle class (Tanesia 2002: 2). Even in regions – like Sumba – which did not have indigenous royalty, a new and 'improved' version of traditional costume was developed for the specific purpose of lavish wedding photos.

3   In India, there is a whole category of costumes where men dress us as gangsters (dacoits), Pathan tribesmen and Arab sheiks. This would seem to parallel Wild West tourist sites such as Virginia City, Nevada, where I have been photographed dressed as an outlaw, or (at the Italian amusement park Gardaland) as a pirate. While Indonesian photo studios do often pose people with (fake) heirloom swords or daggers, I am not familiar with bandit costumes specifically. The costumes rented out by wedding studios are exclusively regional costumes (*pakaian adat*) and royal uniforms, without the hint of transgression that these others seem to include. (MacDougall (1992) even mentions a short-lived fascination with hippie costumes in India.)

Buitenzorg Place (meaning 'free of cares', and at one time a refuge for colonial officers to escape the heat and hardships of their more remote stations), where the Dutch Governor General once lived. A great many Dutch families went there for picnics and weddings (as we know from the colonial photographs in *Tempo Doeloe*) and the Belgian princess Astrid had a honeymoon cottage built for her 1928 visit with her new husband King Leopold of Belgium (Kosasih 2002: 4).

Today, the Botanical Gardens are the setting for some of the most lavish weddings of Indonesia's elite, which include photo sessions for the young couple in the middle of a pond, surrounded by lush tropical vegetation and a Dutch colonial palace. The cost of photo session alone is about 8 million rupiah ($400) for a basic set of twenty photographs, and the use of a series of different costumes (all European, from 'summer dresses' to formal wedding gowns and colonial uniforms). The wealthiest families in Jakarta, both indigenous Indonesian and ethnic Chinese, travel 56 kilometers from their homes to come to be photographed here (Kosasih 2002). The photographs produced are modeled on romantic film posters, and the whole principle of these images is similar to those of the young couples MacDougall observed being photographed in Mussourie: 'For the honeymoon couples who come here tourism is perhaps the ultimate extension of the cinema experience, with themselves, for the only time in their lives, in the leading roles' (MacDougall 1992: 121). This form of photography 'assists in the creation of a reality, not in the discovery (or uncovering) of it' (ibid: 123). And it is especially intriguing that these photographs are essentially of 'natives in white face' – they invert the racial and political order of the colonial period to picture Indonesian elites playing in Western garb in a garden once restricted to Europeans.

Conclusions

Seeing cameras in the hands of natives rather than seeing them only through the other end of an intimidating lens is often disconcerting to present day tourists. Because of the prevalence of the colonial stereotype, people in remote Asian villagers are sometimes assumed to be naïve even when they are technologically sophisticated. Pushpa Tulachan, a doctoral student in our visual anthropology program at USC returned in 2000 to his native Nepal to work on a Nova documentary about the Himalayas. Pushpa was watching a village festival when he was approached by a Western photographer who pointed repeatedly at himself saying, 'Me America, from America', humming a few bars from the 'Star Spangled Banner'. Then the man planted his tripod in front of Pushpa and began an elaborate pantomime, grabbing at the air with his hands and gesticulating to show how images were caught up from the air and went into the metal box erected on legs in front of him. 'Camera', he shouted as he waved his arms, 'not dangerous', although all his movements seem to indicate precisely the opposite. 'Yes', replied Pushpa in virtually perfect English, 'that is a really great Hasselblad you have'. The man's jaw dropped almost to his knees.

This story illustrates how efforts to dispel what Westerners believe to be Eastern fantasies of soul-stealing cameras can end up only spreading strange rumors about the technical processes involved.

A photographic portrait is a person flattened into a thing, a two dimensional likeness. Benjamin sees the early focus on portraits at the dawn of photography as an echo of the idea of capturing an ancestral image: 'The cult of remembrance of loved ones, absent or dead, offers a last refuge for the cult value of the picture. For the last time the aura emanates from the early photographs in the fleeing expression of a human face. This is what constitutes their melancholy, incomparable beauty' (1968: 228).

My argument looks at photographs and film images in the context of ancestor worship and animism, but it moves in exactly the opposite direction of Benjamin's: Rather than seeing the celluloid image as the 'last refuge of the cult value of the picture', I see it instead as the cutting edge of a postcolonial perspective on modernity. By reflecting on who has the power and the resources to produce and especially reproduce images mechanically, indigenous peoples articulate new forms of visuality, and new ways of engaging with the restrictions placed on their own visual culture. They may formulate accounts of enchanted cameras, but they do so not in a state of worshipful awe of Western technological advances, but instead in a stance of critical questioning of inequities in global communication.

While it might have proved a safer strategy to write this chapter as a simple critique of the now rather widely acknowledged Western fantasy of the soul-stealing camera, I have chosen instead to try to focus discussion on an emerging discursive formation which offers, in my view, a more trenchant critique of global inequities in access to the technology of mechanical visuality. Through the distorted mirror held up to Western industrial societies by indigenous media, we can come to see a truth in the apparently most 'primitive' phantasms of cameras as animated and voracious engines of technological domination.

References

Banta, M. and Hinsely, C. (1986), *From Site to Sight: Anthropology, Photography and the Power of Imagery*, Peabody Museum Press, Cambridge.

Barthes, R. (1981), *Camera Lucida: Reflections on Photography*, translated by R. Howard, Hill and Wang, New York.

Benjamin, W. (1968), 'The Work of Art in the Age of Mechanical Reproduction', in H. Arendt (ed.), *Illumination*, Schocken, New York, pp. 217-252.

Berger, J. (1972), *Ways of Seeing*, Penguin, London.

Berger, J. and Mohr J. (1982), *Another Way of Telling*, Pantheon Books, New York.

Broder, P.J. (1990), *Shadows on Glass: The Indian World of Ben Wittick*, Rowman and Littlefield, Savage, MD.

Carpenter, E.S. (1972), *Oh, What a Blow That Phantom Gave Me!*, accessed online at Virtual Snow, <www.faculty.virginia.edu/virtualsnow>. Accessed 17 April 2009. Originally published by Holt, Rinehart and Winston, New York.

Cohn, B. (1989), 'Cloth, Clothes and Colonialism: India in the 19th Century', in A. Weiner and J. Schneider (eds), *Cloth and Human Experience*, The Smithsonian Institution Press, Washington.

Djelantik, A.A.M (1997), *The Birthmark: Memoirs of a Balinese Prince*, Periplus Editions, Singapore.

Dwyer, L. (2002), 'Culture on Camera', *Latitudes Magazine*, August, accessed online at <www.LatitudesMagazine.com>. January 3, 2003.

Edwards, E. (ed.) (1992), *Anthropology and Photography, 1860-1920*, Yale University Press, New Haven and London.

Edwards, E. (2001), *Raw Histories*, Berg, Oxford.

Gell, A. (1999), 'The Technology of Enchantment and the Enchantment of Technology', in E. Hirsch (ed.), *The Art of Anthropology: Essays and Diagrams*, Athlone, London School of Economics Monographs on Social Anthropology, London.

Gillen, F.J. and Spencer B. (1968), *Native Tribes of Central Australia*, Promotheus Books, Sydney.

Griffiths, A. (2002), *Wondrous Difference: Cinema, Anthropology, and Turn-of-the-Century Visual Culture*, Columbia University Press, New York.

Hoskins, J. (1993), *The Play of Time: Kodi Perspectives on Calendars, Exchange and History*, University of California Press, Berkeley, CA.

Hoskins, J. (1996a), '"Letter for an Angel": An Indonesian Film on the Ironies of Modernity in Remote Areas', *Visual Anthropology Review*, Spring, pp. 67-73.

Hoskins, J. (1996b), 'The Heritage of Headhunting: Ritual, Ideology and History on Sumba 1890-1990', in her (ed.), *Headhunting and the Social Imagination in Southeast Asia*, Stanford University Press, Palo Alto, pp. 216-248.

Hoskins, J. (1998), *Biographical Objects: How Things Tell the Story of People's Lives*, Routledge, London.

Hoskins, J. (2002), 'Predatory Voyeurs: Tourists and 'Tribal Violence' in Remote Indonesia', *American Ethnologist*, Vol. 29, No 4, pp. 603-630.

Keane, W. (1994), 'The Value of Words and the Meaning of Things in Eastern Indonesian Exchange', *Man*, Vol. 39, pp. 605-629.

Keane, W. (1997), *Signs of Recognition*, University of California Press, Berkeley, CA.

Keesing, R. (1984), 'Rethinking Mana', *Man*, Vol. 40, pp. 137-156.

Kosasih, U. (2002), 'An Oasis called the Bogor Botanical Gardens', *Latitudes Magazine* November issue, accessed online at <LatitudesMagazine.com>, January 3, 2003.

MacDougall, D. (1992), 'Photo Hierarchicus: Signs and Mirrors in Indian Photography', *Visual Anthropology*, Vol. 5, pp. 113-129.

Mracek, R. (2002), *Engineers of Happy Land: Technology and Nationalism in a Colony*, Princeton University Press, Princeton.

Needham, R. (1976), 'Skulls and Causality', *Man*, Vol. 11, pp. 71-88.

Pinney, C. (1990), 'Classification and Fantasy in the Photographic Construction of Caste and Tribe', *Visual Anthropology*, Vol. 3, pp. 259-288.

Pinney, C. (2002), *Camera Indica: The Social Lives of Photographs in India*, University of Chicago Press, Chicago.

Poole, D. (1997), *Vision, Race and Modernity: A Visual Economy of the Andean Image World*, Princeton University Press, Princeton.

Ryan, J. (1997), *Picturing Empire*, Cambridge University Press, Cambridge.

Sahlins, M. (1994), 'Goodbye to Tristes Tropiques: Ethnography in the Context of Modern World History', in R. Borofsky (ed.), *Assessing Cultural Anthropology*, McGraw Hill, New York.

Sekula, A. (1984), *Photography Against the Grain*, Art Metropole, Los Angeles.

Sekula, A. (1989), 'The Body and the Archive', in R. Bolton (ed.), *The Contest of Meaning: Critical Histories of Photography*, The MIT Press, Cambridge, MA.

Smedal, O. (1989), 'Order and Difference: The Orang Lom of Bangka', *Oslo Papers in Social Anthropology*, Vol. 19.

Sontag, S. (1977), *On Photography*, Delta Books, New York.

Spyer, P. (2001), 'Photography's Framings and Unframings: A Review Article', *Society for Comparative Study of Society and History*, Vol. 10, pp. 181-92.

Stocking, G. (1993), 'The Camera Eye as I Witness: Skeptical Reflections on the "Hidden Messages" of Anthropology and Photography, 1860-1920', *Visual Anthropology*, Vol. 6, pp. 211-218.

Stoler, A. (2002), *Carnal Knowledge and Imperial Power*, University of California Press, Berkeley, CA.

Tanesia, A. (2002), 'Royalty for a Day: Wedding Culture in Java'. *Latitudes Magazine* October issue, accessed online at <LatitudesMagazine.com>, January 3, 2003.

Taussig, M. (1987), *Shamanism, Colonialism and the Wild Man*, University of Chicago Press, Chicago.

Thompson, J. (1874), *Illustrations of China and its People*, Geographical Society, London.

Turner, T. (2002), 'Representation, Politics, and Cultural Imagination in Indigenous Video: General Points and Kayapo Examples', in F. Ginsburg. L. Abu-Lughod and B. Larkin (eds), *Media Worlds: Anthropology on New Terrain*, University of California Press, California.

Wachtel, N. (1994), *Gods and Vampires: Return to Chipaya*, translated by C. Volk, University of Chicago Press, Chicago.

White, L. (2000), *Speaking with Vampires; Rumors and History in Colonial Africa*, University of California Press, Berkeley, CA.

Chapter 9

Re-Viewing the Past: Discourse and Power in Images of Prehistory[1]

Andy Letcher, Jenny Blain, and Robert J. Wallis

Introduction

The prehistoric stone circles of Britain, Stonehenge in particular, have achieved iconic status in contemporary culture. Images of them appear on postcards, in advertisements, in newspaper cartoons and repeatedly in television documentaries, to the extent that stone circles have become the obligatory, if clichéd, signifiers of Britain's 'ancient past'. Yet these representations might merit only passing academic attention – or perhaps analysis as British tourist attractions for an international market – were it not that stone circles are sites of contested meanings, and in the case of Stonehenge, often bitterly so. Archaeologists, heritage managers, tourists, religious adherents of Paganisms and 'alternative spiritualities', earth mystics, 'new-age travellers', and festival-goers, all express very different interpretations of stone circles, interpretations which generate incompatible modes of engaging with them. Inevitably some interpretations and modes are legitimised, whilst others are marginalized or even criminalized.

Given that these varied groups produce and consume images of prehistoric sites, we might expect to see these diverging positions variously represented pictorially in a variety of ways. Indeed this is the case; far from being 'neutral', images of stone circles are discursive devices used to articulate, negotiate, and contest the meanings that these sites convey. This chapter introduces the variety of representations of stone circles, principally of Stonehenge, and through discourse analysis, reveals underlying and often obscured positionings of their authors. For instance, the clichéd dramatic images of the sun rising behind the deserted silhouette of Stonehenge, are indicative of a particular view of how the Stones should be physically encountered, managed, and interpreted: they serve to promote a discourse of the past favoured and legitimated by many heritage managers. The chapter interrogates our (modern Western) engagement with and re-presentation of the past by questioning such images and (connotative or denotative) narratives produced by heritage managers, tourists, and 'alternative'

1   The authors wish to acknowledge the support of the ESRC through grant RES-000-22-0074 which enabled preparation of this chapter.

interest groups. Finally, we argue that images of Stonehenge have created 'hyper-real' expectations amongst 'visitors' which the physical reality cannot meet, and how plans for a new visitor centre are designed, in part, to compensate for this. We draw on previous work of the Sacred Sites, Contested Rights/Rites project (see <www.sacredsites.org.uk>) and on critical discourse analysis informed by the work of Wodak and Meyer (2001), Chouliaraki and Fairclough (1999) Potter and Wetherall (1987), Smith (1990), and elsewhere developed by Blain (see, for example, 2002). It is not 'neutral': members of the project have their particular relations to sites, groups, Pagan spiritualities and to heritage management, which are made explicit in the project's work and elsewhere (see, for example, papers in Blain, Ezzy and Harvey 2004)

More fully: our project adopts methods of reflexive ethnography/auto-ethnography and visual anthropology, together with discursive and textual analysis as described above. We are examining and considering not only how materials (images or text) are presented, but the concepts and ideas that are associated with that presentation. In our using of critical discourse and visual analysis, images or text cannot be void of social/political positioning, and the juxtaposition of text and image in the various documents (heritage management or Pagan) that we are scrutinizing, has impacts both intentional and unintentional on diverse communities of readership and use.

Anthropology has been critiqued for, at the same time, an intense reliance on visual material as documentation, together with a disregard for the means of production of that material and its locatedness, and indeed a kind of 'iconophobia', a fear of images as, perhaps, too 'seductive, dazzling, deceptive and illusory ... regarded as capable of wreaking all sorts of havoc with the sobriety of the discipline' (Grimshaw 2001: 5). In choosing to inflict a textual/discursive analysis on photographic imagery of 'sacred sites', our Sacred Sites project – a collaboration between an anthropologist (Blain), an archaeologist (Wallis), and, in this chapter and during the period of our ESRC funding, a researcher from Religious Studies (Letcher) – is adopting the position that our own locations within academia and with respect to Paganisms are crucial to the problematising of meaning that follows. In short, somebody else would do it differently. The reflexive methodologies we adopt necessitate examination of our own locations and interactions with sacred sites. In this chapter we focus on Stonehenge, with its high public profile and tourist associations: as an 'icon of Britishness' it also is a key site for many Pagans, with its own recent history of contest and appropriation (for example, Pendragon and Stone 2003). Other sites for which we are following Pagan use and heritage discourse include, in England, Avebury, Stanton Moor, Thornborough Henges, Stanton Drew, Ilkley Moor, Castlerigg, and various in the Peak District and Cotswolds; and in Scotland, Argyll's Kilmartin Glen.

Archaeology and the Preservation Ethos

The dominant, legitimized discourse pertaining to public prehistoric sites is that articulated by those groups entrusted with their management: archaeologists and heritage managers. This so-called 'preservation ethos' rests on the foundation that the past is 'closed', and whilst it may be knowable (interpreted) to a greater or lesser extent, the past is ultimately unreachable. Thus archaeologists interested in the religious significance of Stonehenge look to the archaeological record, perhaps even to historical records, but not to the present day. Contemporary ritual activity is invalid as a form of archaeological knowledge, can not be considered 'real', because it lies outside of this arbitrarily defined boundary of closure (see also Finn 1997). What counts as being 'of interest' are the ritual activities of the 'original' culture(s).

The past may be closed and unreachable, but within the preservation ethos it holds 'value', both intangible and tangible. Its tangible value is partly that it provides the material objects, the physical evidence, which enables archaeologists to interpret the past (and have a profession), and partly economic: through tourism the past generates significant amounts of revenue.[2] However, the preservation ethos also sees the past and its artefacts as possessing an intangible value, in and of themselves, that must by necessity be preserved: 'the remains of the past, both moveable and immovable and also including the intangible – the historic environment – should be protected and preserved for their own sake and for the benefit of future generations' (APPAG 2003: 6). Rather like a Victorian butterfly collection, the now lifeless relics from the past are to be pinned down, labelled, protected and conserved so that future generations of people can also look at and treasure them. It is an ethos, therefore, that privileges intellectual and visual engagement with the past and its material objects, and archaeologists as final arbiters of meaning (such heritage discourse is also examined critically by, for example, Lowenthal 1996; Bender 1998).

Important implications follow from the routine application of the preservation ethos to the management of stone circles, implications which are particularly evident at Stonehenge. First, the ethos is primarily concerned with the activities of people in the past as 'scientific' knowledge, and the impact that these activities may have had upon a site, to the detriment of contemporary 'non-scientific' activities and any resulting amelioration. Given that the ethos demands that a site must be preserved in its current state, and that change constitutes 'degradation', contemporary activities must necessarily be regarded as potentially threatening

---

2 It is important to see this as a political positioning. Archaeology, and 'heritage' more generally, is chronically under funded, and archaeologists little esteemed – despite the television construction of 'media stars' within the profession. The construction of 'heritage' is a means to create public interest, and as the APPAG meeting in December 2002 made plain, public involvement is essential to legitimation of archaeology as knowledge within today's Britain, in which everything has its price.

a site's stability and continuity. In other words the ethos creates the situation in which sites are regarded as being continually and permanently under threat. Thus Stonehenge is 'under pressure not only as a result of the large numbers of visitors that it attracts, but also from the impacts of traffic … [p]ost-war agricultural intensification, changes in military use … and increasing demands for leisure and recreational use of the countryside' (English Heritage 2000: 1.3). Management plans seek to minimise 'visitor pressure', to determine a site's 'carrying capacity'[3] or 'sustainable' number of visitors, and to agree 'limits of acceptable change': 'The physical environment at "the Stones" has proved unable to withstand pressure from the number of visitors with the result that strict visitor management measures have been introduced such as roping off the Stones' (English Heritage 2000: 3.22). From the point of view of the preservation ethos, therefore, visitors to Stonehenge and other sites constitute a fundamental problem.

The discursive nature of this position is made clear when the history of Stonehenge is examined. Stonehenge ('monument' and 'landscape') has never been a singular thing but is rather a palimpsest, or composite, resulting from many phases of intervention. Throughout its long history the monument itself has been altered, changed, developed, abandoned, and reused (see Chippindale 1994; Bender 1998). The name, Stonehenge, dates to the Anglo-Saxon period, some 2000 years after it was constructed – and is itself of uncertain derivation. The layout of the monument with which we are familiar dates back only to the 1950s, when stones fallen since the eighteenth century were re-erected, the rest left as examples of 'damage' from ancient times. Thus the felt need to preserve Stonehenge as it currently exists is a comparatively recent development, and one which foregrounds a prehistoric 'ruin' apparently frozen in time.

Second, the ethos privileges a visual, rather than physical or tactile, mode of engagement with sites: to appreciate sites fully they should be *looked* at, like museum exhibits, but preferably from a distance so that they can be appreciated in the full context of their archaeological landscape (though not all archaeologists support this view without question: see, for example. Thomas 1993; Bender 1998; Wallis 2003). Management plans for Stonehenge repeatedly stress the importance of landscape, and distance views. Stonehenge is 'not so much a site as a "cultural landscape"; a relict landscape preserving evidence of a long history of man's interaction with the environment' (English Heritage 2000: 2.4). 'Typically much of the WHS is an open landscape in which the sky dominates' (English Heritage 2000: 2.15). 'From the highest points on the downland (often the location of the most

---

3   A term borrowed from environmental discourse, in which it means something slightly different. In mathematical modelling the carrying capacity is the stable population that an environment can support. The use of the term in heritage management to suggest the maximum number of visitors a site can support without damage to the site is a subtle discursive shift. In ecology the environment constrains the population and not vice versa. ('sustainability' is similarly so derived).

significant archaeological remains), views are extensive and landscape features can be visible from a considerable distance' (English Heritage 2000: 3.13).

The visual bias inherent within the preservation ethos is consequently influential upon current archaeological interpretations which foreground spatiality, distance and perspective, and vice versa. Furthermore it has contoured thinking regarding the planned improvements to visitor facilities at Stonehenge, and the proposed 'visitor experience'. Plans supported by English Heritage and the Highways Agency were (pre-2008) to replace the busy A303 with a four-lane highway, whether via a bored tunnel through and beneath the environs of the monument or by re-routing the road;[4] the A344 and the current visitor centre are to be closed and turfed over (or in part painted green!) and a new visitor centre built some several miles from the Stones. Visitors will be bussed to vantage points from which to view the panorama of 'the landscape' (now perceived to be untrammelled by recent human architecture), and thereby to appreciate the perspective, apparently, of archaeologists. It will constitute the latest (and likely not the last) amelioration of the landscape in the history of Stonehenge. What people will see is *Stonehenge 201X* (it is unlikely that any plan will begin to be implemented before the second decade of the twenty-first century), but whether this is actually what visitors want remains a moot point (Baxter and Chippindale 2002; see below).

Unsurprisingly therefore, this discursive position may be found in the images of Stonehenge produced by heritage managers and archaeologists. These representations typically show Stonehenge from a variety of dramatic, unusual, and long distance perspectives (i.e. from the air, silhouetted against the sunrise, and so on), reinforcing the notion that the grandeur and 'mystery' of the place can only be experienced from a distance. The Stonehenge Project web site, www. stonehengeproject.org.uk, is a case in point. Moreover, and in contrast to most people's experience of the site, they invariably show a Stonehenge devoid of humanity. The Stones are rendered not as architecture but as 'natural', framed as part of a dramatic landscape in which humans no longer play a part. The Stones appear frozen in time, hence they are 'timeless', 'unchanging', and ultimately 'mysterious' (see, for example, on-site presentation panels at Stonehenge visitor centre); qualities set in stone which heritage managers have committed themselves to preserving.

---

4   Road plans were contested by various heritage and conservation agencies, including the National Trust, CPRE (Council for the Protection of Rural England) and ICOMOS-UK (International Council on Monuments and Sites, UK), and the Pagan-led ASLaN (Ancient Sacred Landscape Network); tunnel plans were withdrawn as too costly and possibilities for rerouting the A303 were problematic; other plans are still current. A history of this particular contested landscape is summarized in our volume from the Sacred Sites Project (Blain and Wallis 2007). Details updates and critiques regarding roads and visitor centre can be found at the Save Stonehenge website (<www.savestonehenge.org.uk>) and current road plans are available on the Highways Agency site (<www.highways.gov.uk/roads/projects/3659.aspx>)

Additionally, archaeologists regularly feature in documentaries (discussed further below) which, whatever the positive educational results, tend to reinforce the preservation ethos through clichéd images and sounds. Archaeologists, as 'experts', are seen striding purposely through deserted sites, or through virtual tours of 'stone-age' Britain, as it is currently imagined. A surfeit of digital editing shows stones 'looming' from unusual camera angles, shrouded in dry-ice, and accompanied by a soundtrack of stereotyped vox-effect drones (contrast with equivalent representations of cathedrals). Archaeologists, eager to promote their discipline, use documentaries to forward their most recent interpretations and, indirectly, the preservation ethos. However, in doing so, and in conceding to the demands of 'good' television, that is, with high ratings, these programmes tend to portray stone circles primarily as mysterious, enchanted places. The archaeological message is thereby obscured by the visual trickery, a romanticisation which makes both the programmes and the sites themselves all the more appealing to wider publics. It is ironic, therefore, that the images and representations of these sites produced by heritage experts are instrumental in generating the very tourism which they believe threatens sites' stability and continuity (see, for example, Lowenthal 1996). It is to a consideration of tourism, and the meanings of sites to non-specialist visitors, that we now turn.

Stonehenge and Tourism

'Stonehenge is perceived internationally as a "must see attraction"' (English Heritage 2000: 2.11), but for heritage managers wedded to the preservation ethos, the increasing numbers of visitors to Stonehenge (up to 800,000 per annum) are problematic. On the one hand visitors contribute to the site's tangible value, in that tourism generates considerable revenue, but on the other, visitors present one of the identified threats to the stable continuity of the stones through the steady erosion of so many feet. At Avebury, for example, tourists are reluctantly accepted by the National Trust: the management plan states that the 'purposes of the National Trust embrace the principle of providing public access, where it is consistent with its primary duties of preservation and good neighbourliness. The Operational Guidelines for nomination to the World Cultural Heritage list do not, in fact, require sites to be accessible to the public' (National Trust 1997: 25). The plan is actually committed to reducing visitor numbers; its idealised image of the site, we argue, is exemplified by an engraving from 1724, in which the site is deserted, unnoticed except by antiquarians. The plan, in other words, side-steps the issue of why stone circles have become so popular, or why they have become so meaningful within contemporary culture, in a reactionary reification of (unrealistic) 'no-impact' engagement.[5]

---

5 Space does not permit us to elaborate on the recent history of Avebury, particularly Alexander Keiller's attempts to reconstruct prehistoric Avebury, involving re-erecting

Interest in the past is currently booming. Channel 4's successful archaeological programme, *Time Team* (first broadcast in 1994), commands an audience of 5 million, whilst the unexpected popularity of his *History of Britain* (first broadcast in 2000) propelled Simon Schama to the status of celebrity academic. Schedulers have responded to this audience-led demand with a plethora of documentaries covering all periods: history and archaeology are the surprise successes of recent broadcasting. The past, and particularly the distant past, *matters* to people.

The reasons why this is so are complex and beyond the scope of this chapter. However two points can be mentioned briefly. The first is that the people distant in time or space, at the 'periphery', provide the 'other' upon which contemporary hopes and fears of those at the 'centre' (in the present) can be projected (for discussion of this important issue across disciplines see, for example, Piggott 1968; Said 1978; Bowman 1995). During times of confidence and faith in the 'progress' of mainstream culture, the other is represented as exotic and primitive. At times when people become disillusioned with the mainstream, the exotic other becomes desirable, 'primitives' become 'noble savages'. The second is that during times of rapid social change, the past represents a point of continuity and stability: hence 'traditions', implying continuity, may be resurrected or invented (Hobsbawm and Ranger 1983). That we are moving both through a time of rapid social change and disillusionment with traditional sources of authority is generally agreed by scholars (see Heelas et al. 1996; Sardar 2002), and consequently the past has become increasingly important as a stable reference of identity and meaning. The great (physical) longevity of Stonehenge means that it, quite literally, represents continuity with the ancient past – or with *an* ancient past, the past in the imagination of the present – which is regarded as unchanging. Furthermore, its inherently 'unknowable qualities' make this distant past particularly amenable to polysemic interpretation: Stonehenge can support a host of alternative meanings in a way that, for example, a cathedral, with a known history, can not. For contemporary culture, destabilised by the uncertainties of post-modernity, the appeal of Stonehenge is considerable.

Additionally, Stonehenge has been given an even greater international significance through its inclusion (along with its 'sister-site' Avebury) as a 'World Heritage Site'. The inscription of these modern 'wonders of the world' has attributed to them an 'outstanding universal value' (English Heritage 2000: 1.2) for all of 'humanity'. It has also delineated these sites as icons of national identity. Thus the Taj Mahal is iconic of, and epitomises, 'Indianess', the Great Wall of China 'Chineseness', and Stonehenge and Avebury variously 'Englishness',

---

stones and removing villagers and their homes. The current role of the National Trust at Avebury is contentious; bad-feeling endures on the part of villagers, and increasing pagan interests, including among locals who are pagans, pressing for tactile engagement (including celebrations for some days at the times of pagan festivals) present new challenges to the preservation ethos. For further discussion on this and other issues addressed by the Sacred Sites project, see Blain and Wallis 2007.

Figure 9.1   Idealised presentation of a Stonehenge in isolation
Source: © Sacred Sites Project

'Britishness', even 'Celticness' or some other variant of constructed (and also contested) UK identity. Foreign visitors to the UK go to places such as Stonehenge (alongside the Tower of London and Madam Tussauds) not just to experience the past, but also to partake of this quintessential branded English/British experience.

Visitors to Stonehenge therefore find meanings in the site that may or may not accord with the preservation ethos. Whilst some may appreciate the site intellectually for its archaeological interest, the majority seek some kind of *emotional* encounter with it, with its past or with its significance as national icon (see Baxter and Chippindale 2002). As Baxter and Chippindale have argued, the majority of visitors do not want to appreciate the Stones from a distance as part of a landscape, but close at hand (ibid). They want to experience exactly those qualities of drama and mystery they have been primed to expect through the plethora of stereotyped images to which they have been exposed. They want to photograph themselves against the background of the Stones so that on their return home they can say 'I have been there'. For Baxter and Chippindale, therefore, the new plans for visiting Stonehenge represent the imposition of a peculiar archaeological discourse on how the site should be interpreted and encountered. We might add that and this discourse is driven, more specifically, by the concerns of heritage management, if not heritage marketing. It is a discourse shaped by the concerns of supply, not demand.

Alternative Meanings

The past is appealing to visitors, then, because it creates a sense of stability or continuity. However, for many in the post-modern world, the past represents a standard against which contemporary life can be critiqued. The past represents a 'golden age' in which some essential lack of modern times can be met. For many contemporary people that lack is one of enchantment – the imagined past re-enchants. The variety of Pagan spiritualities, their diverse forms united by the commonality of looking to the past for sources of belief, practice and identity, are some of the fastest growing contemporary religious movements. Similarly, Earth Mystics and Alternative Archaeologists seek patterns of meaning in the past 'overlooked', 'ignored' or simply not considered applicable by mainstream archaeology. A typical, widely held, belief by Pagans and Earth Mystics alike is that 'ancient peoples were sensitive to aspects of the physical world that are not appreciated in modern day society' (Lambrick 2001: 20).[6]

Practitioners exhibit 'topophilia' (Bowman 2000), and thus sites such as Stonehenge are not simply for them places of archaeological interest. For many they are sacred places, temples even, which appear to hold answers to important questions of meaning and ontology, and to meet the need for re-enchantment. However, if, as Baxter and Chippindale contend, visitor expectations and meanings are largely ignored, then those of Pagans, alternative archaeologists, earth mystics, travellers and free-festivallers are actively marginalized and have even been criminalized. At an abstract level, alternative discourses are incompatible with, and directly challenge, the preservation ethos. For alternative practitioners the past is not closed, nor is it unreachable; its relics are 'living' things. Thus many consider Stonehenge to be a living temple, as sacred now as it was to the 'ancestors'. Through a variety of practices, which include dowsing, meditation, visualisation, ritual, entheogen consumption, and the cartographic search for 'ancient alignments', it is believed that the knowledge of these ancestors can be rediscovered, and the lack in our culture met. It is understandable why academic archaeology, resting upon rational enquiry and the procedures of graduate training, peer review and so on, is suspicious of these alternatives, for there is a fundamental methodological and epistemological incompatibility.

However, it follows from alternative discourses that intellectual engagement with sites from a distance is not sufficient, and the need to interact physically, emotionally and spiritually with sites is what renders these discourses so problematic

---

6   This discourse parallels, and in part replicates, attempts by indigenous people in North and South America, Australia and elsewhere, to claim knowledge of landscape and place: it is an inherently political discourse of the twentieth and twenty-first centuries, through which indigenous groups have been able to advance (for instance) land-claims and so re-claim some degree of self-determination. This political dimension is often not recognised in 'new age', English Heritage, or Pagan usage. Nevertheless, the political implications remain.

Figure 9.2   Stonehenge experienced at summer solstice, 2001
Source: © Sacred Sites Project

for the preservation ethos. That managers perceive alternative visitors as a threat is evidenced by the following statement, unsubstantiated with data: 'Visitor pressure is compounded by the highly seasonal nature of tourism at Stonehenge together with peaks created by the influx of visitors at certain times of the year, mainly at the summer and winter solstices and equinoxes' (English Heritage 2000: 3.22). Elsewhere [in *The Times*], bank holidays are seen as the time when site is most threatened). The Stonehenge management plan does attempt to accommodate these alternative views to some extent, saying 'it is essential that plans for visitor management do not affect the essential mystical quality of Stonehenge, a quality which for many people lies at the root of its attraction' (English Heritage 2000: 3.23). It is possible to negotiate an 'hour in the stones' for groups, at a price, and while strictly controlled, these private Druidic or other group engagements, are tolerated and/or negotiated with (Blain and Wallis 2002). Druidic and other Pagan rituals are now so commonplace as to be represented on postcards at Avebury, and advertised in literature produced by the Red Lion pub. However, rituals and alternative views are only accommodated to the extent that the preservation of the site is not compromised – the preservation ethos remains non-negotiable.

The situation has been made more complicated by the emergence, growth, and eventual suppression of the Stonehenge free festival. Running for ten years from 1974-84, the festival, with its overt contravention of societal conventions, drug laws, and with its black market economy (Hetherington 1992), unsettled the authorities; its proximity to the archaeology of Stonehenge, in turn, unsettled archaeologists. Its demise was inevitable. What the authorities underestimated was how important the festival had become to those attending. Not only did people construct significant aspects of their identity and lifestyle around it, but also, through its utopian vision and entheogen consumption, it met the fundamental need for re-enchantment. People were attached to the festival in a very deep, sometimes religious, sense, and would act to further its continuity. The forced closure of the festival, the resulting confrontations, and the demonisation of travellers and festival-goers, merely served to give Stonehenge a powerful new meaning (see, for example, Bender 1998; Worthington 2004, 2005). It became an icon of freedom and alterity for anyone dissatisfied with Thatcher's Britain (McKay 1996). Going to Stonehenge, or at least making the attempt, became an act of resistance.

More recently a lengthy negotiation process has taken place between heritage personnel, Pagans and other celebrants (on recent developments see Chippindale et al. 1990; Sebastion 2001; Worthington 2004; Blain and Wallis 2004, 2007), resulting in the managed opening of Stonehenge over the solstices, allowing free access to all comers within a designated period (the resurrection of the free-festival, whilst the goal of many alternative interest groups involved in negotiation, seems unlikely to be tolerated within the vicinity of the Stones in the foreseeable future). The negotiation is not without its problems, and we present the following example, of disagreements over the appropriate day to celebrate the winter solstice, to illustrate the complexities resulting from these differently held discursive positions. Not only does there exist contestation/negotiation between adherents of alternative and preservation discourses, but also amongst alternative practitioners.

In 2001, according to many Pagans, the correct moment at which to observe the winter solstice at Stonehenge was the morning of Friday 21 December, and a number of Druids, travellers, and others determined to be at Stonehenge then. Other Pagans, however had already negotiated access with English Heritage for the following morning, Saturday 22 December, and subsequently English Heritage had agreed that Stonehenge would only be open on the Saturday morning: any people demanding access on the Friday morning would be turned away (Everard, pers. com.). In effect, whatever their intentions to facilitate a negotiated managed access, English Heritage were seen as dictating to Pagans and other celebrants when their festival days should be. As anticipated, a considerable number of people turned out at Stonehenge on 21 December (more, in fact, than on the 'official' day) and, as stipulated, English Heritage initially denied access. However as verbal confrontations between celebrants and security guards broke out, the police prompted the guards to open the gate. Those present, who had not trespassed at this point, now deemed they had been invited into the stones and the proceedings thereafter were peaceful.

Feedback after the event – on various email lists, in person to various people involved, and gained from various conferences we have attended – raises further points pertinent to our discussion over forms of access and engagement with Stonehenge. Clearly, English Heritage were concerned that events had not proceeded as they, as site managers, would have liked. In a telephone interview, Clews Everard (then director of Stonehenge) suggested that 'criminal damage' had been caused to gain entry with an electric fence and a padlock broken; further, glass bottles were left behind and people climbed on the stones 'with no regard for the sanctity of the place' (pers. com.). She added that the years of work negotiating access were compromised by these latest events, leaving English Heritage in the position of having to review how this would affect the future. In other words English Heritage felt that its role as 'preserver' of the Stones had been compromised.

Press reports the following week are also revealing, the description of the 'unofficial' Friday event – 'Revellers break into Stonehenge ... THREE hundred revellers broke a padlock in order to greet the dawning of the shortest day of the year ... in defiance of an order from English Heritage to stay away until the following day' (<www.thisiswiltshire.co.uk> 27 December 2001) – contrasting with the description of the 'official' event which was so 'peaceful' the headline could say no more than 'Winter solstice is an all-weather event ... Druids ... were treated to snowflakes, a rainbow and a clear sunrise' (<www.thisiswiltshire.co.uk> 28 December 2001).

A similar situation occurred at the winter solstice 2002, and, more crucially because of the large numbers potentially involved, the summer solstice 2003, but without any anticipated confrontation emerging, though a considerable amount of negotiation seems to have been employed in the latter case, and some people were being 'turned away' and attempting to go elsewhere on the day they considered to be the 'true' solstice. What these disputes – ongoing at the time of writing in the run-up to summer solstice 2007 – reveal are the underlying tensions that exist between incompatible discourses pertaining to sites. From our viewpoint as observers and participants, it is exactly these issues of 'ownership' (of knowledge or interpretation, rather than physical ownership of the monument) that are central to all 'sides' in this debate.

Stonehenge and Hyper-Reality

Druids, Pagans, festival-seekers and other 'alternative' visitors typically find the experience of Stonehenge falling far below their expectations. Indeed the image of the Stones has been so over-represented that visitors of all shades, particularly tourists expecting hyper-reality, find reality a disappointing second. Fenced off, ringed by throngs of other visitors and watchful attendants, set off against two heavily used roads, Stonehenge-as-actually-experienced is very different from the images of dramatic and mysterious solitude. No wonder Stonehenge has been famously, repeatedly described as a 'national disgrace'.

Our final point, then, concerns the future of Stonehenge and the proposed management plans which will see the creation of a new 'visitor experience'. These plans have been shaped in large part by the need to fill the 'reality gap'. However, rather than shape the representations to fit reality, the landscape is to be managed so as to make it fit more closely with visitor expectations (and with the needs of preservation) – as announced by the Stonehenge Project website. The planned removal of the A303, A344, and the relocation of the visitor centre can, we argue, be read as cosmetic landscaping exercises to contour the land to fit unrealistic expectations of a 'real' prehistoric landscape in situ. Visitors can now expect to see the long-distance views and dramatic panoramas for themselves. Baxter and Chippindale have cast serious doubt on the feasibility of a greenfield site on which all can wander, pointing out that however fashionable landscape archaeology has become, it is the stones themselves that people want to see, touch, experience at first hand. Indeed, the ability to wander amongst the Stones and encounter their mysterious qualities first hand, initially promised by the landscaping plans, appears to have been since withdrawn. The current text of an FAQ on the Stonehenge Project Website responds to the question 'Will I be able to go into the Stone Circle?' by explaining:

> Access inside the stone circle has been stopped since 1978 because of erosion, vandalism and increased visitor numbers. However, it is possible to book private access outside opening hours. In the future, these arrangements will remain. (<www.thestonehengeproject.org/news/faq7.shtml>).

However, in case this customized on-site reality is still not enough, the new visitor centre is to contain the latest virtual reality technology, so that hyper-real expectations can be met. It may be hoped, by those charged with preserving the site, that the lure of virtuality coupled with a bus ride and a mile walk will deter many from venturing to the Stones. We will all have to wait and see whether this is indeed the outcome.

Conclusion

It is clear from this discussion that Stonehenge is not a singular thing. Stonehenge is polysemic, in that it signifies a range of meanings, discursively contested through image and text. Our position here (and with the Sacred Sites project more generally) has not been to favour any particular interpretation, but rather to reveal the often hidden positionings of image-makers and consumers, in order to further dialogue. Given that all images, regardless of context, tend to show Stonehenge as it is imagined, progress might be made if there was a turn towards producing images of Stonehenge as it is experienced by visitors at the current time – a diversity of henges, to display a multivocality of meanings. To do that, however, would be to undermine the very potent and almost universal need for Stonehenge to remain 'essentially preserved', shrouded in mystery, the ancient guardian of a hidden past. Such an entrenched perspective will not be surrendered lightly.

References

APPAG (All-Party Parliamentary Archaeology Group) (2003), *The Current State of Archaeology in the United Kingdom*, The Caxton and Holmsdale Press Ltd, Sevenoaks.

Baxter, I. and Chippindale, C. (2002), *A Sustainable and Green Approach to Stonehenge Visitation: The 'Brownfield' Option* (Discussion Paper: pdf file available on request; contact info@moffatcentre.com).

Bender, B. (1998), *Stonehenge: Making Space*, Berg, Oxford.

Blain, J. (2002), *Nine Worlds of Seid-Magic: Ecstasy and Neo-shamanism in North European Paganism*, Routledge, London.

Blain, J., Ezzy, D. and Harvey, G. (2004), *Researching Paganisms*, Altamira, Walnut Creek.

Blain, J. and Wallis, R. (2002), *Heritage, Paganisms and a Climate of 'Transparency': Autoarchaeological Method and the Sacred Sites, Contested Rites/Rights Project*, paper presented at the 5th Cambridge Heritage Seminar.

Blain, J. and Wallis, R. (2004), 'Sacred Sites, Contested Rites/Rights: Contemporary Pagan Engagements with the Past', *Journal of Material Culture*, Vol. 9, No 3, pp. 237-261.

Blain, J. and Wallis, R. (2007), *Sacred Sites, Contested Rites/Rights: Pagan Engagements with Prehistoric Monuments*, Sussex Academic Press, Brighton.

Bowman, M. (2005), 'The Noble Savage and the Global Village: Cultural Evolution in New Age and Neo-Pagan Thought', *Journal of Contemporary Religion*, Vol. 10, No 2, pp. 139-149.

Chippindale, C. (1994), *Stonehenge Complete*, Thames and Hudson, London.

Chippindale, C., Devereux, P., Fowler, P., Jones, R. and Sebastion T. (1990), *Who Owns Stonehenge?*, Batsford, Manchester.

Chouliaraki, L. and Fairclough N. (1999), *Discourse in Late Modernity: Rethinking Critical Discourse Analysis*, Edinburgh University Press, Edinburgh.

English Heritage (2000), *Stonehenge World Heritage Site. Management Plan*, accessed online at <http://www.english-heritage.org.uk/>.

Finn, C. (1997), 'Leaving More Than Footprints: Modern Votive Offerings At Chaco Canyon Prehistoric Site', *Antiquity*, Vol. 71, pp. 169-178.

Grimshaw, A. (2001), *The Ethnographer's Eye: Ways of Seeing in Modern Anthropology*, Cambridge University Press, Cambridge.

Heelas, P., Lash, S. and Morris, P. (eds) (1996), *Detraditionalization*, Blackwell, Oxford.

Hetherington, K. (1992), 'Stonehenge and its Festival – Spaces of Consumption', in R. Shields (ed.), *Lifestyle Shopping – the Subject of Consumption*, Routledge, London, pp. 83-98.

Hetherington, K. (2000), *New Age Travellers: Vanloads of Uproarious Humanity*, Routledge, London.

Hobsbawm, E. and Ranger, T. (eds) (1983), *The Invention of Tradition*, Cambridge University Press, Cambridge.

Lambrick G. (2001), *The Rollright Stones: Conservation and Management Plan 2001-2005*, The Rollright Trust, Banbury, Oxfordshire.

Lowenthal, D. (1996), *Possessed by the Past: The Heritage Crusade and the Spoils of History*, Free Press, New York.

McKay, G. (1996), *Senseless Acts of Beauty: Cultures of Resistance Since the Sixties*, Verso, London.

National Trust (1997), *Avebury Management Plan*, National Trust, London.

Pendragon, A. and Stone, C.J. (2003) *The Trials of Arthur*, Element, London.

Piggott, S. (1968), *The Druids*, Thames and Hudson, London.

Potter, J. and Wetherall M. (1987), *Discourse and Social Psychology: Beyond Attitudes and Behaviour*, Sage, London.

Said, E. (1978), *Orientalism: Western Constructions of the Orient*, Penguin, Harmondsworth.

Sardar, Z. (2002), *The A to Z of Postmodern Life*, Vision, London.

Sebastion, T. (2001), 'Alternative Archaeology: Has it Happened?, in R. Wallis and K. Lymer (eds), *A Permeability of Boundaries? New Approaches to the Archaeology of Art, Religion and Folklore*, British Archaeological Reports, Oxford, pp. 125-135.

Smith, D. (1990), *Texts, Facts and Femininity: Exploring the Relations of Ruling*, Routledge, London.

Thomas, J. (1993), 'The Politics of Vision and the Archaeologies of Landscape in Barbara Bender (ed.), *Landscape: Politics and Perspectives*, Berg, Oxford, pp. 19-48.

Wallis, R. (2003), *Shamans/neo-Shamans: Ecstasy, Alternative Archaeologies and Contemporary Pagans*, Routledge, London.

Wodak, R. and Meyer M. (2001), *Methods of Critical Discourse Analysis*, Sage, London.

Worthington, A. (2004), *Stonehenge: Celebration and Subversion*, Heart of Albion Press, Loughborough.

Worthington, A. (ed.) (2005), *The Battle of the Beanfield*, Enabler Publications, Teignmouth.

Chapter 10

Entwined Histories: Photography and Tourism at the Great Barrier Reef

Celmara Pocock

Introduction[1]

From the air a scattering of brilliant white sandy cays and sand fringed islands dotted in an array of navy-black, brilliant aqua and turquoise waters stretches for more than 2,000 kilometres along the northeast coast of Australia. Up close the waters are crystal clear, the islands are green, and underwater life displays colours and forms unimaginable to those who have never seen it. This is the Great Barrier Reef – simultaneously of enormous scale and comprised of myriads of minute life forms. It defies the human imagination, but satellite and aerial imagery, underwater cameras, colour emulsion, digital technology and motion film make it possible to capture and communicate many of these visual qualities. Images of the Reef are reproduced in their thousands each year; in popular science magazines, documentary films, coffee table books, internet sites, tourist brochures, advertisements and postcards. The visual qualities transmitted through these media are an integral part of the region's standing as a World Heritage site and as Australia's premier tourist destination. While the tourism industry is often attributed with creating and promulgating particular images of tourist destinations, analysis of historic images of the Reef suggests that this relationship is a much more complex one.

Reef images have developed through a number of technological advances, and Reef and photographic histories are intertwined. The history of tourism on the Great Barrier Reef is documented in a vast array of images including many photographs. A content and semiotic analysis of these images, together with other archival sources has been used to identify change in visitor experiences of the Reef over time (Pocock 2002a). This analysis suggests that the physical environments that surround tourist facilities of Reef islands have been transformed to meet the imaginary Pacific ideal realised in many resorts today. This transformation is fuelled not only by the tourism industry, but through the collective historical imagination of colonial Australians. Beyond the islands, visitor experiences of the Reef are strongly influenced by the visual imagery of aerial photography and underwater cameras. This chapter explores a history of these photographic images

1   The research for this chapter was supported and assisted by the CRC Reef, James Cook University, Marion Stell, David Collett, Shelley Greer and David Roe.

and reveals how photography has not only created the Great Barrier Reef as an object for tourist consumption, but also influences the experiences of being at the Reef as a tourist.

A Bird's-Eye-View

The Great Barrier Reef first entered European consciousness through the accounts of Captain James Cook, the intrepid English navigator who sighted the northeast Australian coast in 1770. For weeks the *Endeavour* sailed within a labyrinth of shallow waters, trapped by the raging seas of the Outer Reef and entangled in the calm shoals of the inner lagoon. Captain Cook's successful navigation of these complex waters is understood within a context of his skill and successes in the South Seas and the legacy of this voyage persists in contemporary appreciation of the Reef. Photographic imagery used to promote and perpetuate tourists' love of this region takes many of its cues from this earliest of accounts. In particular, the scale and complexity of the Reef and the exotic qualities of South Sea Islands have both had lasting influences in visitor experiences of the region and in the way it is presented in photographic imagery (Pocock 2003).

As a navigator Captain Cook had the primary task of mapping the lands of the new world. He experienced particular difficulty in navigating the Australian coast because of the apparently endless labyrinth of shoals, reefs and islands along the north-eastern seaboard. Maps and charts created the cartographic control necessary to navigate the complex myriad of hidden dangers that were real threats to early navigators and which continue to ground ships in modern times. Cartography is constructed as though the observer is situated perpendicularly to the depicted region (Ryan 1996). This is a strategic view that facilitates control (cf. de Certeau 1984) and, in the case of the Reef, maps control hidden dangers. So long before the invention of human flight, the bird's eye view was made possible through cartographic imagination. This view not only facilitates control, but encapsulates the many fragments that constitute the region into a single frame.

Captain Cook and subsequent navigators created charts and maps that pieced the many reefs, coral cays, shoals and islands together into a singular whole that was later coined the Great Barrier Reef. Through maps the many single reefs, islands and shoals – or at least a significant proportion of them – can be viewed as a singular Great Barrier Reef. This is significant because the notion of a single and enormous entity is fundamental to the World Heritage status of the region. It is this characteristic that defines the Great Barrier Reef as unique and of outstanding universal value. In other words it is size that sets this reef system apart from other coral reefs of the world. It is also this recognition that underlies its significance as an international tourist destination.

Advances in technology have enabled this strategic and all-encompassing view to be reproduced through aerial and satellite imagery. This is no longer an imagined viewpoint, but one that is attributed with a sense of reality through the power of photography (cf. Sontag 1973; Taussig 1993). These images give rise to the 'unparalleled aerial vista' that is cited as a unique characteristic of the Great Barrier Reef World Heritage Area (Environment Australia). It also facilitates human appreciation of the region as sublime – a source of both fear and inspiration – at least in part because the strategic view simultaneously tames hidden dangers and makes the enormity of the region apparent.

The endless quality of the reefs and shoals is now synonymous with the wonder of the region, but for eighteenth and nineteenth century navigators this was the cause of great fear and frustration. One of the greatest imperatives for navigators was to determine a safe passage for vessels. By climbing to hilltops they could gain the strategic view that allowed them to negotiate a way through the labyrinth of corals. Early tourists also climbed hilltops, though their motivation was more often in quest of scenic panoramas of surrounding seas and islands. However, both navigators and tourists could only access this vantage point from the heights of continental islands. These remnants of a submerged coastline lie close to the Australian mainland and although surrounded by colourful and rich fringing reefs are not considered authentic parts of the Great Barrier Reef. It is coral cays that are regarded as 'true coral islands' and authentic Reef experiences are built around these and the Outer Barrier Reef which lies some distance from the mainland. Consequently, an aerial view of the authentic distant reefs and coral cays and islands is only accessible through cartography or from the air. Such views were not available to early visitors and even today are primarily available through photography.

Full-colour aerial views are regarded as an accessible and accepted part of contemporary experiences of the Reef and are an important tourist commodity. The aerial experience is heavily marketed to travellers and aerial images proliferate in brochures, postcards and web sites. However, few physical encounters are actually perceived from this vantage point. There are opportunities for visitors to go sky-diving or take helicopter or seaplane rides over the Reef. Many visitors travel to, or between Reef destinations by air. To varying degrees, these activities offer bird's eye views of the Reef. For the main part, however, experiences of this kind remain out of the financial reach or desire of many visitors. And for those who do glimpse the Reef from the air, this is only ever one of many holiday experiences.

The capacity of aerial photography to render the Reef a singular entity requires much greater height than most scenic flights offer. This is a view best encapsulated from off the planet. The Reef is often cited as the only living organism visible from space, and it is satellite imagery that brings the impressive scale of the Reef back to earth. This is partly because it is at such a distance that it is possible to measure it against continental landmarks and hence reaffirm its extensive size. Such distant observation is the domain of very few individuals, and it is primarily through photography that most acquire this experience. So it is from a great distance that the mass of reefs can be understood as a single Reef, rather than through an embodied encounter. In other words it is through photography that most people conceive of the aerial vistas and singular nature of the Reef.

Strategic aerial vision of the Reef is therefore achieved through highly sophisticated photographic technology rather than through embodied human experience. Nevertheless the unprecedented aerial views dominate discourse about the beauty of the Reef and its heritage values, and the strategic vista is taken as an integral part of tourist experiences. In practice, the attainment of this vision is predominantly through the postcards tourists purchase and often send away, and from the imagery in brochures and film footage that draws them to the region. In this sense, the aerial and strategic view of the Reef is one that is equally accessible to people at the Reef and those a considerable distance away.

The Living Reef

In contrast with a conceptual single Reef, embodied or physical interaction is confined to particular regions of the World Heritage Area. These regions are themselves comprised of many microcosms. Reef life is diverse, colourful and intriguing. Much of it is also relatively small in scale. Some is microscopic. Such visual detail and diversity is in marked contrast to the broad sweep of aerial vision. Nevertheless such minutia has equally strong currency in contemporary tourism and is an essential component of an authentic Reef experience. It is these life-forms that have captured the imaginations of visitors and would-be visitors throughout the twentieth century. Like aerial imagery, however, views of the underwater world have not always been as accessible as they are today. The role of photography has been a crucial part of tourist access to this sphere.

Photographic technologies are intimately linked to scientific investigation and interpretation (Taussig 1993), and this relationship is extended to Reef tourism which is based in a history of scientific research. Scientific interest in the living Reef dates to the beginning of the nineteenth century. Tourism itself emerged through the participation of holidaymakers in these scientific expeditions in the 1920s. It is also scientific interpretations that underpin the World Heritage status of the region. An analysis of photographic images suggests that Reef photography was primarily developed and refined in order to advance and communicate these scientific findings and interpretations. Tourist interactions with the Reef are therefore influenced by historical traditions in both photography and science.

Underwater Exclusion

Initial research interest in the Reef was economic as well as scientific as is illustrated by the work of naturalist William Saville-Kent (Love 2000: 99). *The Great Barrier Reef: Its Products and Potentialities* (Saville-Kent 1893) places economic exploitation in a context of scientific theory and description of the coral reefs. It is beautifully illustrated by the author's own coloured drawings and superb black and white photographic plates. The photographs take advantage of exceptionally still weather conditions to allow the living corals to be portrayed

Figure 10.1   William Saville-Kent photograph of a reef at Palm Island
Source: Saville-Kent, W. (1893)

from the surface through the tranquil water (Figure 10.1). Saville-Kent stood with a tripod in shallow waters of the Reef to capture these images, and it would have taken enormous discipline and patience to maintain the calm water while doing this. The compositions are also very carefully framed and composed. Without colour Saville-Kent was dependent on the light and shadow of form and texture to illustrate the diversity of life in the coral pools. The deliberateness of these images together with his detailed descriptions allowed later investigators to determine the exact locations captured by the images (Harrison 1997: 104). Love (2000: 103) has suggested that these photographs misrepresent the Reef by framing such perfect conditions and beautiful compositions because few visitors see the Reef like this. She therefore suggests that Saville-Kent's imagery was scientifically misleading and represents an early foray into hyper-reality. However, photographic technology available to Saville-Kent and other early photographers demanded this kind of construction for the living reef to be effectively portrayed.

Early photography was limited in its capacity to communicate aspects of the living Reef, but some of these restrictions also characterised personal experiences. At the beginning of the twentieth century, scientists and holidaymakers had only limited access to the underwater world. They did not have the kind of physical or visual access that is taken for granted today. Early visitors viewed the living Reef by peering through clear water left in shallow coral pools on the exposed reef at low tide. In these vignettes they observed corals and fishes. Holidaymakers and scientists gazed into these small worlds and observed the life within them. So like the camera, people relied on the right weather conditions, low tides and their own patience and stillness to see the Reef. Only in such exceptional conditions could they peer beneath the surface. The innovation of the water telescope or 'waterscope' – a paraffin tin or bucket with its base replaced with glass –eliminated surface ripples and allowed people to view the underwater in a greater range of weather conditions. Like the later glass-bottomed boat visitors could also see Reef life in deeper water and they were less reliant on low tides. However, early visitor experiences remained voyeuristic. The view gained through these technologies was constructed from above the surface and from a perpendicular angle. In other words it was the view of an outsider.

This way of observing the Reef was characteristic of both scientific research and holiday activities. Science was an important aspect of early Reef holidays which were either associated with scientific expeditions, or organised and guided by amateur and professional naturalists. A major part of the scientific work was to collect and catalogue various species. Visitors assisted in these collections, but also gathered their own coral and shell specimens. These were gathered as souvenirs and were particularly important because of the limitations of photography. Cameras could not operate beneath the water surface and rapid movement was unable to be captured on film. Consequently creatures were taken from the water in order to view them in detail. It was also usual to photograph animals in this way, and many images show dead or dying creatures. Thus both photography and personal experience were restricted in similar ways. Both personal experience and photography were constructed from the surface, and intimate views of reef specimens were of creatures out of their natural habitat.

The Problem of Colour

While there are some similarities between these photographic representations and bodily experiences, some aspects of personal engagement were beyond the scope of early photographic technology. This is particularly noticeable in relation to colour. The underwater Reef has always been a focus of tourist fascination, and a predominant theme in both contemporary and historic visitor accounts relates to its luminous and striking colours. Early visitors raved about the brilliance and diversity of colourings and markings of fishes and corals. In spite of their numerous adjectives and lengthy accounts, many authors resort to describing the colours as 'indescribable' and 'beyond the human imagination' (Pocock 2002b). This suggests that in the first part of the twentieth century neither words nor photographs were sufficient to communicate the experience of being at the Reef itself.

Bringing these experiences to a distant place remained a real challenge for much of the twentieth century. The difficulty of communicating the living Reef was a major impetus for the development of aquaria which have the potential to represent living creatures in close proximity without the loss of colour and movement. However, many early attempts to transport tropical creatures were unsuccessful (Pocock 2002c). Colour had to be represented by other means. Saville-Kent had experimented with triple-layered glass transparencies in an attempt to convey the colours of the corals (Harrison 1997: 95), but it was his paintings that were reproduced as hand-coloured plates in his book. Later photographers hand-tinted their black and white landscapes with pale pink or yellow sunsets and painted coloured edges onto black and white clam lips. Postcards were also hand-coloured, and bleached white coral displays were painted to represent living colours. But the paint boxes of the late nineteenth and early twentieth century could do no justice to the brilliance of the coral reef. In promoting her own skill as a coral painter, Shirley Keong wrote in 1965 that 'displays of coral should not be of ... "icing sugar" colourings' (Prime Minister's Department 1965-1966). However, where colour was represented at all, it tended to very pale. This can be seen in one of the earliest published coloured photographs which appeared on the front cover of the October issue of *Life* Magazine in 1933 (Purcell 1933). The image appears muted and pale in comparison with contemporary images of Reef scenery.

Underwater Participation

The *Life* Magazine cover is not only limited by the depth of colour it portrays. It also represents the Reef from a surface or outsider perspective that characterises many images of this time. The image is not of underwater corals and fishes, but rather shows a group of fishers aboard a boat. While photographic technology was incapable of representing the underwater world effectively, this was also the case for human interactions with the living Reef. Very few people physically encountered the underwater world until the second part of the twentieth century. One of the few who did was Mel Ward, an enthusiastic naturalist. He donned a pair of eye goggles so that he could plunge himself below the surface and gain a participant's view of the Reef. Ward accompanied several Embury Expeditions in the 1930s. These expeditions were some of the first organised group visits to the Reef designed especially for holidaymakers. They were organised around a central party of scientific researchers who would lecture and guide people on Reef life. The organiser, Mont Embury, was supported by two professional photographers; his brother, Arch Embury, and Otto Webb. They took many photographs of visitors and scenery and these images were reproduced in Australian and international newspapers, journals and magazines, including *National Geographic Magazine*. Arch Embury recalled that during one of the Embury expeditions they conducted early experiments with underwater photography:

> On one occasion Mel, Mont and I carried out some of the earliest Australian attempts at Underwater Photography. Mel intended going on a Lecture Tour in America and required underwater Reef Pool shots for his slide series. I got together what was one of the earliest underwater cameras by building a plate glass fronted case, inside which my camera was fitted with lens pointing through the glass front. Our method of operation had Mel swimming around the bottom of a pool probing amongst the rocks with his prospectors pick and on one occasion wrestling with a turtle! I leaned out over the pool, poked the glass front

Figure 10.2   Underwater photograph of Mel Ward at the Great Barrier Reef, 1930

Source: Arch Embury © Mitchell Library, State Library of NSW, Australia.

> of the case under the surface and worked the shutter, while Mont's job was to hang back onto a strap around my waist to prevent me going headfirst into the pool with Mel. We got quite a good series of pictures. (Arch Embury, March 1981 (Mitchell Library 1925-1945))

The images from these experiments were published in newspapers, magazines and brochures. One that was reproduced several times, including in *The World* in October 1932, shows Mel Ward diving underwater (Figure 10.2). Although this is not a high quality image and shows little of the living Reef, it is significant in that it shows a view from within the underwater sphere. This is possibly the earliest underwater image of the Reef and through it the underwater was made accessible and available to a large audience for the first time.

This participant's view was not commonly available to early holidaymakers themselves. Underwater observatories were built on Green and Hook Islands in the 1950s and these gave visitors the opportunity to view the Reef from below the surface. They did not require visitors to get into the water. While people swam in netted enclosures on the island beaches and used the sea to bathe, diving to view the Reef was uncommon until the late 1960s. Instead it was underwater photography that brought the perspective of immersion to the surface. The majority of visitors continued to view the Reef by staring into coral pools or peering through a waterscope or glass-bottomed boat. Like the aerial images of the Reef, it was through photography that tourists gained their first side-on views of the underwater world. In other words it is through photography that the iconic views of the Reef, the aerial vistas and colourful detail, have been achieved and through which people have come to understand their own experiences.

Towards a Complete Photography

Gradually during the twentieth century visitors gained greater access to the underwater world through underwater viewing chambers, snorkelling and scuba-diving equipment. At the same time underwater cameras, motion film and colour emulsions were developed. Reef holidaymakers were quick to take advantage of these new developments and by the end of the century rich colour, motion film and underwater cameras were not only the preserve of professional photographers, but were in the hands of nearly every visitor. Like early photographic technologies new advances were adopted by both scientific researchers and tourists alike. This allowed tourists to emulate scientific documentary makers and to engage in the activity of capturing this experience themselves. Taking photographs and making films was therefore as important as seeing the Reef for themselves. Even early descriptions of underwater life suggest that watching fishes in coral pools is analogous to watching photographic imagery:

> You take a handful of bread and mix it with a handful of chopped meat, and you take it with you to the coral garden. You adjust your water telescope – this is usually an ordinary dipper with a glass bottom cemented in. When it is placed on top of the water and you look through the glass there is no ripple to obstruct the view, and you can see everything below as clearly as you can in your room at home. You make yourself comfortable by sitting down in the six inches or so of water at the edge of a pool, and place your feet on the coral ledge below. Then you begin to 'listen in'. You throw your bread crumbs and meat into the pool, and your 'picture show' commences. ("Whampoa" 1930)

The analogy with motion film is one that has continuing implications for the way in which the underwater world is experienced today. Later developments in motion film and video enabled sound as well as visual amenity to be captured. Taussig (1993) suggests that modern film allows us to approach something of the sensate eye – the camera moving as one in flight. Contemporary experiences of the Reef are constructed within this visual sensuousness. The bodily experiences of being at the Reef are removed from the landscapes and associated smells, tastes and sensations but they take on their own novel forms (Pocock 2002c, 2003). Floating free of particular location, the body is weightless and because of this visual experience is more three dimensional than on land. The experiences are slightly disembodied, the visual heightened because of a loss of usual bodily sensations. Sound too is strangely focussed on the immediate body and light is somewhat surreal.

The sense of a free-floating body is also approached through aerial imagery. Thus both macroscopic and microscopic views of the Reef are constructed from above, below and from the side. These two ways of viewing the Reef are brought together in many photographic contexts such as brochures, books and documentaries. In this way the aerial view is juxtaposed with the intimacy of a single coral. However the dual visual experiences of the Reef from out of space and the proximity of underwater engagement can only be experienced simultaneously through photographs. Similarly the capacity of photography to replicate colours of the underwater Reef has now surpassed the personal encounter. Night cameras allow corals to be filmed at their most extended. So while few visitors can or want to dive at night, the intense colour of the night dive is a standard image portrayed in magazines and postcards. Again this is not the way that the majority of tourists experience the Reef, but it nevertheless fulfils their perception of what the Reef offers visually. Unlike the images of Saville-Kent that allowed people to return to exact locations, today's images are interchangeable with one another and the precise location of the observer is irrelevant to the visual experience. In this way photography has heightened the way in which the Reef is experienced and suggests an experience of hyper-reality.

Photography as Reef Practice

Photography is dominant in Reef experience and knowledge and plays a central role in its construction and visitor experiences of the region. The reasons for this are multiple and overlapping. Explanations are not simply found in a visually dominant tourism industry, though that undoubtedly is highly influential. Tourists at the Reef seek to replicate the images that dominate tourism brochures (Albers and James 1988; Watts 2000), but also to engage with the Reef in a way that is constituted through European navigation and scientific enquiry.

As visitors once collected their own vast array of shells and corals, prolific photography fulfils this scientific tradition. Conservation requires that visitors do not plunder the resources of the Reef as they might have in the past. However, as visitors have been excluded from these kinds of interactions, photographic technology has become increasingly sophisticated and accessible. It has thus been able to effectively replace earlier activities of fossicking, collecting and cataloguing. The effectiveness of this displacement relates to the link between photography and science and the way in which photography maintains, replicates and enhances aspects of the original. Photography at the Reef has historic continuity that parallels the scientific tradition of collecting. This gives it currency as both a tradition and a science. In the second instance photography plays an important role in maintaining the essence of copy and contact that underscores the significance of the collecting tradition. The power of shell and coral collections lay in their status as part of the Reef itself. Through the perception that photography retains some element of the original (Sontag 1973; Taussig 1993) images of the Reef can similarly maintain this physical link with the Reef itself. In other words, photography has been able to replicate historical continuity in scientific method and the essence of contact between the original and copy. For this reason people continue to visit the Reef and take their own photographs. Even though visitors have access to a vast array of professional images their personal photographs capture something of themselves. The importance of image does not relate to its artistic quality (Sontag 1973), but on the act of taking the image and thus of being a part of the image. For these reasons photography has proliferated as an important Reef activity, possibly beyond that of other tourist destinations.

Conclusion

In this chapter I have suggested that the primary visual paradigms through which the Reef is understood and experienced by visitors; the aerial vistas and the underwater world, are largely brought to all people by the camera. In some instances technology has been unable to represent visual experiences, but in many others it has produced new ways of conceiving the Reef. Furthermore, some experiences created by the camera are only possible through these technologies. The juxtapositioning of close-up and distant views represents an instance in which photographic simulacra create visual experiences that do not reflect embodied and personal encounters with the region (cf. Baudrillard 1983). Photography also facilitates a somewhat contradictory conception of the Reef as both a single entity visible through satellite photography and the minutia of a single coral polyp made possible through microscopic underwater photography. The simultaneous presentation of these two views conflates any part of a single reef with the whole Great Barrier Reef. The use of sophisticated lighting, satellite technology, lens filters and night photography have all heightened visual representations of aerial vistas and underwater life. These technologies produce much brighter, clearer and intimate views of the living corals and the single Reef than is possible with the human eye. It is this heightened visual sense of the Reef that shapes both visitor and would-be visitor expectations and experiences and creates a sense of hyper-reality in Reef experiences (cf. Eco 1986).

Photographic imagery and photographic technology are integral to the way in which tourists and would-be tourists experience the Reef. In these images any microcosm of a living reef acts as a synecdoche of the singular Great Barrier Reef which is encapsulated and experienced through aerial photography. The images brought to us are brighter and clearer than our own vision. They can simultaneously present us with distant and close up imagery in a way that is not possible in person, and which nevertheless heightens our sense of the Great Barrier Reef.

References

Albers, P.C. and James, W.R. (1988), 'Travel photography: a Methodological Approach', *Annals of Tourism Research*, Vol. 15, pp. 134-138.

Baudrillard, J. (1983), *Simulations: Foreign Agents Series*, Semiotext(e), New York.

de Certeau, M. (1984), *The Practice of Everyday Life*, University of California Press, Berkeley and Los Angeles.

Eco, U. (1986), *Travels in Hyper-reality*, Picador, London.

Environment Australia (2002), *Great Barrier Reef World Heritage Values*, <www. environment.gov.au/heritage/places/world/great-barrier-reed/values.html>, accessed 17 April 2009.

Harrison, A.J. (1997), *Savant of the Australian Seas: William Saville-Kent (1845-1908) and Australian Fisheries*, Tasmanian Historical Research Association, Hobart.

Love, R. (2000), *Reefscape: Reflections on the Great Barrier Reef*, Allen & Unwin, St Leonards, Australia.

Mitchell Library (1925-1945), 'Rough copy of letter written to Marion Mahon by Arch Embury March 1981', in *Embury Scientific and Holiday Expeditions on the Great Barrier Reef, Mitchell Library, State Library of New South Wales* PXA 642, 187 Sydney.

Pocock, C. (2002a), 'Identifying social values in archival sources: change, continuity and invention in tourist experiences of the Great Barrier Reef', in V. Gomes, T. Pinto and L. das Neves (eds), *The Changing Coast*, Eurocoast/ EUCC, Porto, pp. 281-290.

Pocock, C. (2002b), 'Sense matters: aesthetic values of the Great Barrier Reef', *International Journal of Heritage Studies*, Vol. 8, No. 4.

Pocock, C. (2002c), 'Through the looking glass: control and colonisation of the Great Barrier Reef', in J. Turtianinen, M. Lähde and E. Kokkoniemi-Haapanen (eds), *Crossroads in Cultural Studies: Fourth International Conference*, University of Tampere, Tampere, Finland, p. 117.

Pocock, C. (2003), *Romancing the Reef: History, Heritage and the Hyper-real*, Ph.D Thesis, James Cook University.

Prime Minister's Department (1965-1966), 'EXPO 67 – Great Barrier Reef Exhibit', in *National Archives of Australia (National Office): A463/50; 1965/4559* Canberra.

Purcell, H.A. (1933), 'With rod and reel on the Barrier Reef', *Life*, October 14.

Ryan, S. (1996), *The Cartographic Eye: How Explorers Saw Australia*, Cambridge University Press, Cambridge.

Saville-Kent, W. (1893), *The Great Barrier Reef of Australia: Its Products and Potentialities*, W.H. Allen & Co., London.

Sontag, S. (1973), *On Photography*, Farrar, Straus and Giroux, New York.

Taussig, M. (1993), *Mimesis and Alterity: A Particular History of the Senses*, Routledge, New York and London.

The World (1932), 'Exploring the wonders of the Great Barrier Reef', in *The World*, 17 October.

Watts, J.A. (2000), 'Picture taking in paradise: Los Angeles and the creation of regional identity, 1880-1920', *History of Photography*, Vol. 24, No. 3, pp. 243-250.

"Whampoa" (1930), 'The Great Barrier Reef: living flowers', *Bank Notes*, August, 20-23.

Chapter 11

The Embodiment of Sociability through the Tourist Camera

Joyce Hsiu-yen Yeh

Introduction

Drawing on fieldwork in Cambridge and London and post-tour interviews in Taiwan, this chapter examines Taiwanese tourists' photographic acts and how these tourists use the camera and photography to engage and perform sociability while they are traveling and when they return home. It begins with an examination of functions of the tourist camera. The theoretical framework chosen for these analyses draws on a wide range of writers engaged in the critique of visual culture and representations through images that have cultural significance in shaping touristic experiences. This chapter integrates theories of identity, social interaction and representation by focusing on the role of the camera in staging tourist performance. I suggest that the use of photography is not only a way to record what tourists see, but also an act to construct social relationships among the tourists and with others. The moment of picture taking affects the social relationship of the photographer to the photographed and can be used to construct a variety of desired personal relationships in numerous settings. Why is it important for most tourists to travel with a camera and to take pictures? How does photographic acting contribute to a sense of group solidarity and act as a way to encounter others? How does photography shape tourists' experiences and constitute their performances? How can we read these activities as forms of social communicative interaction?

Functions of the Tourist Camera

Academic interest in tourism, visual culture and photography has grown in recently years. For example, Richard Chalfen's work (1979), in which he approaches the unexplored connections between photography and tourism, gives some examples of various anthropological attitudes towards tourists' 'shooting' the subjects and objects of the host culture to remind us of the issue of appropriate camera use by tourists. In a similar fashion, Albert and James offer a useful systematic methodological approach for examining the relationships between photography and tourism (1988: 134-158). In their research on travel photography, they acknowledge that one of the tourist acts is, for some tourists, to take photographs exactly like those of postcards or advertising brochures. They argue that photographs like postcards provide a medium through which an effectively grounded aesthetic can be shown and enable personal experiences to be shared. It is important to note that the accessibility and mobility of a camera makes photography and tourism inseparable. Martin Parr (1995) provides many examples of such 'inseparability' between the tourist and the photographic act, which has become another of the 'photographic' sights that attracts the lens of both tourists and professional photographers. Kevin Markwell comments on this phenomenon of photographic practices, suggesting that to be a tourist 'is to be, almost by necessity, a photographer' (1997: 131). It can be argued if the act of photographing can be seen as an identifiable marker of being a tourist, it can also be used to make a distinction of such labelling by rejecting of the act. Not all tourists take photographs. Some 'travellers', especially, avoid showing their cameras so they are able to perform a 'non-tourist' behaviour and distinguish themselves from mass tourists. However, what is significant about this 'non-tourist' act is the very distinctive conception of the camera as a maker of tourist identity.

The use of camera and the materiality of images have played a major role in representational being and sense of aesthetics. Discussing aesthetics and ethics in *Distinction* (1984), Bourdieu shows how the photograph is an example of a means of the accumulation of cultural capital (1984: 44-47). In *Photography: A Middle-Brow Art* (1990b), Bourdieu not only argues that photography is a form of souvenir, he also indicates the social function of photography. As he points out, 'the need to take photographs can be understood as a need for photography' (1990b: 25). Here, photography becomes especially important because it is intimately tied to the expression of tourist identity and on the social aspects of the performance of photographic practices. And yet, what is missing is the analysis of the need for tourists to take pictures and the significance and meanings of accumulating photographs, in which particular cultural characteristics should be taken into account. As I have argued elsewhere (Yeh, 2009), despite the rapid growth of Asian tourism and tourist practices, Asian tourists' photographs remain largely unexplored and few empirical studies have addressed the issue from the perspective of dialectical encounters between the East and the West. In this chapter I am interested in the relationships between the tourist-as-photographer and tourist-as-photographed as well as the ways in which the photographic act and tourists' travel photographs can be seen as embodiments and performances of sociability. To explore this relation further, I now turn to a discussion of the functions of the camera within the context of tourism where touristic images are made, collected and circulated.

In the next section of this chapter I consider the essential apparatus of photography – the camera. The role of the camera will be used as a focus, investigating its dynamic relationship with tourists. Why does the camera require a separate discussion? Precisely because of its role in understanding tourist encounters and because not enough sociological research has been done on the role of the camera in constructing tourist performance. Annette Kuhn suggests, 'with the right equipment to hand, we will make our own memories, capture all those moments we will some day want to treasure, call to mind, tell stories about' (1995: 25). The functions of a camera are a fundamental issue in understanding the role of photography and its relavion with tourism, but so far this issue has not been tackled in a sustained and empirical way (Chalfen 1979). A camera is linked with tourists in various ways. A camera, as Chalfen states, is a tourist's 'identity badge' for them to 'do tourism' (1979: 436). This identity badge signals the agency of the tourist-as-outsider. While I support Chalfen's recognition of the essentiality of the camera to tourists, cameras are also necessary visual tools for tourists to demonstrate their cultural and collective identities in their touristic performing acts.

The photographic act, Kevin Markwell points out, is an 'important social activity for tourists serving to strengthen bonds among fellow tourists' (1997: 135). All the Taiwanese tourists in this study travelled with their cameras. Cameras inevitably bring a shared identity into the group's space and establish the web of relations within the group. The structure of the network enables the members of the group to comprehend instantly that agreement on the photographic act is recognized and accepted. With cameras, study tourists know what to expect within the group, as everyone wants photographs of their holiday. This same desire provides a group of unrelated individuals with a channel for breaking the personal and social boundaries and legitimizes the immediacy of body contact. This involves kinaesthetic movements among the group members. I observed study tourists at times when they took group photos. In order to put everyone in the frame they put their arms around each other in order to shorten the distance between them. Sometimes, they made gestures, such as 'V', behind or above the person next to them to create the 'funny' effect in the photographs. Sometimes, they 'argued' with each other about which poses or positions were better. They might have 6 to 10 group photos taken at the same spot from various positions but always with the same group of people. In this way, group members can transcend distance and build up a closer relationship with others. Consequently, cameras enhance the intensity of contacts in the group, bring a group of tourists together and alter their relationships from strangers to familiar in-group members who can create and share memories and perhaps become friends later.

Being a study tour leader and travelling with a group of study-tourists, I witnessed how a non-human agency, the camera, alters group dynamics. Cameras are a tourist's performing tool and are signifiers of Taiwanese study-tourists' kindred-ness and uniformity within a group of former strangers. This group met at the airport for the first time and started travelling together as strangers, but the photographic act lessened the distance between them. Every study tourist whom I observed and spoke to had the experience of asking other group members to take a picture for him/her in a place of 'Englishness'. They offered or asked for help from each other to take individual pictures of themselves and the process of negotiation of how the picture was to be taken broke the social barrier between them. Through exchange of each other's cameras to have their own pictures taken, tourists position themselves in a network of relations to others and allow themselves to be connected to the 'closeness' of a given moment. I witnessed how they talked, inquired, negotiated then smiled for the camera so they could produce the picture that they wanted. Study tourists helped, cooperated and sometimes argued with each other to shape the view they wanted and complete the photographic act. Gradually, they established much closer relationships than those who did not participate in their photographic activities. I also noticed that the process of picture-taking was getting longer and longer during the later weeks as they spent more time in discussions and negotiations between the photographed and the photographer: where to point the camera, what should be included in the frame, who to invite to join in the picture, and even how to pose. The embodiment of sociability is reflected in how the proximity of space among study tourists can demonstrate the solidity of a group. The group photo legitimizes shortening physical distance and acquiring the acceptance of the sense of 'closeness' with each other. Through photographic acts, tourists develop active and desirable relationships with the other group members. It transforms the strangeness among the group members into a developing familiarity while in England.

The camera is a universal communicator as well. Tourists' cameras contribute to the creation of a connection between self and others outside the group. For example, a camera can alter the social interaction between strangers and tourists (see Cohen and Almogor 1992). With their photographic act tourists extend the connections with others. One study tourist put it this way:

> I like to ask strangers to take pictures for me. It gives me the chance to talk to them...Without my camera, I don't think I would dare approach strangers like that. In fact, most of them are very nice. Some of them show their interest in knowing where I come from. So we start having a conversation. (Sylvia,[1] a 17-year-old high school student; 1 August 2000)

It is the camera that generates the interaction between the tourists and strangers. The act of taking photographs of someone, or offering to take a photo of a group, may serve as a signal to initiate a conversation and bring about a personal encounter with others. This account also suggests the tourist's strategy for finding a 'proper' mode of social interaction between the self and the other. The desire to encounter 'strangeness' and exchange conversations indicates the urge to meet others and the photographic act achieves this goal. In this context, the camera is a social tool to build up relationships with others and generate the social interaction that study tourists seek.

The above account also reflects certain photographic practices that the Taiwanese study tourists perform. Many of them tend to have many self-portraits taken within

---

1   All the English names were chosen by the study tourists in this study. All the interviews were done in Chinese and the Chinese transcriptions the English translations had been sent to the interviewees to avoid the misunderstood or misinterpreted.

various settings and enjoy being photographed in the sites of 'otherness'. Another study tourist put it more forcefully:

> The camera is a must. I always travel with my camera. When I'm abroad I like taking many photographs including myself in the settings. And I guess that's why my photos are different from the postcards. Postcards might have better pictures than my photos but I wasn't there…My own experience of looking at my family's or friends' photo albums is exactly the same. I skip all the buildings or landscape shots and only focus on people who are in the pictures. And I can tell you that my friends are the same. (15 September 2001)

This account suggests that the tourist self can never be described without the reference of Otherness, which is displayed in the same frame. Taking a great number of self-snapshots is a way of relating to the site of 'Otherness'. Tourists use cameras to enact this linkage between the self and 'Otherness' and mediating 'Englishness' in order to take home photographs as self-made souvenirs, so that people at home are able to follow their gazing at Englishness. The consequence of linkage of the self and the Other to the relational context is the power of the camera.

Donald Horne in *The Public Culture* (1986) suggests the link between the power of a camera and the creation of 'reality' and argues that photographers or tourists have the choice to express their own subjective 'truth'. He writes:

> When we hold cameras in our hands we hold an enormous potential to be artists, to create 'reality' in our own way. Yet, as if by choice, almost all of us use this potential to portray existence in exactly the same manner as everyone else. Although, with these ritual acts of stereotyping, we can feel we have choice (it is we who are making the cameras click), it is as if we are using these small black boxes to demonstrate our faith that there is only one truth. (Horne 1986: 1)

What Horne demonstrates is the ambiguity between individual and collective photographic practices. On the one hand, the camera is a means of expressing individuality and a sense of 'creativity'. On the other hand, the tourist seems only to capture clichéd images. Despite the result that tourists' photographs may be identical to those of other tourists, the moment of the photographic act is very individual and subjective. Once the photograph has been taken, the tourists present a kind of individual 'creation' of the 'reality' that belongs to how they see and sense the world around them.

Everything can be framed and constructed through the lens of a camera. With the camera, tourists take an active role in creating pictures and in reshaping the meanings of received sites that they hope will highlight their travelling experiences. The camera empowers the tourist's ability to produce and to mobilize cultures. This should not be taken as merely entailing a social construction of images, but as allowing new forms of cultural images to develop. Each travel photograph demonstrates the tourist's way of gazing and his/her photographic practices. Urry has argued 'there is no single tourist gaze as such'(2002: 1). Indeed, it is difficult to talk of the tourist gaze in the singular and understand the study tourists' cultural and social experiences as a homogenous group. Urry (1990a) has argued that there is a power/knowledge relationship between see-er and seen. For Urry, the 'tourist gaze' is structured and constructed and is highly socially designed. In particular, tourist photography, therefore, is a 'socially constructed way of seeing and recording' (1990a: 138). In accord with Sontag's argument, Urry further explains the significant meaning of photography to tourists. He writes:

> Photography gives shape to travel. It is the reason for stopping, to take (snap) a photograph, and then to move on. Photography involves obligations. People feel that they must not miss seeing particular scenes since otherwise the photo-opportunities will be missed. (Urry 1990a: 139)

The gaze depends on tourists' viewpoints and on what they understand of their visit as well as the 'reality' that they want to reflect on. It is also noticed that the meaning of what is gazed at might change, but what people choose to look at is also in part socially determined. How does a camera, 'an extension of the eye', relate to the 'tourist gaze' as Urry (1990a and 2002) put it? Who directs the gaze? The photographer can exercise his/her power to control how to see and what to see through the act of photography. In other words, the photographer guides the 'tourist gaze' and frames the view. Between the framed and the framing of the view for photography, the question arises as to who has been seen and what is to be seen. In other words, photographic conversations shape the way photos are taken.

The frequent use of photography also raises the relation with what Sontag called 'power'.

> To photograph is to appreciate the thing photographed. It means putting oneself into a certain relation to the world that feels like knowledge – and, therefore, like power. (Sontag 1979: 4)

Indeed, the act of photography is also an act of power, the 'power' to exclude the unwanted image of Englishness and to include the most 'desirable' one. Photograph-taking exercises some control of the world (see Maxwell 1999). Tourists seem to have choice and power as they use the camera to document their touristic experiences and choose different 'factual' recordings in the images. After the tour, tourists enact their power through interpreting their travel photographs. In this sense, cameras and photographs are vehicles to perform tourists' 'will' and 'choice' of 'authenticity' of the modern world, as MacCannell (1973; 1976) claims.

As travel with a camera is a common phenomenon for tourists, it also intensifies the anxiety to take pictures. The camera as a tool can be a burden too, especially when it does not belong to the tourists themselves. A16-year-old high school student, Jessica, said:

> My parents lent me their expensive camera so I can take good pictures to share with them. Although I appreciate their kindness, I also found it's a huge burden to take care of it. Another burden is that I have to take as many pictures as I can because they want to see everything. As my father said, that camera could record my life here so they can have a better idea what I've been doing. (26 July 2000)

Despite the responsibility of taking care of the camera, the burden also comes from the anxiety and ambiguity of taking pictures for parents, particularly in relation to 'trouble' shots which might have not been approved by their parents. Jessica's comment also reveals different uses of the camera and photographs for her parents. The camera, in this sense, traces the ways in which study tourists' lives away from home/parents may also be seen in part as of long-distance parental surveillance. Drawing upon the legacy of Foucault (1977), study tourists' cameras are forms of parental 'governance' and entail greater social regulations of control over study tourists' behaviour when they are abroad.

The anxiety to have as many pictures taken as possible is common among Taiwanese study tourists. Nick, a 17-year-old high school student, told me that he took the same shots as the other study tourists because he did not want to miss anything. He said:

> I don't know much about Britain so I photographed everything, especially those buildings and churches. They looked impressive although sometimes I have no idea what buildings I've visited or seen. I guess I can check the guidebooks when I go back to Taiwan but now I just want to shoot everything...I also follow other members, if they took pictures I did the same thing too. (14 August 1999)

Using the camera to take pictures is the way in which tourists concretise their gaze at Englishness and make it into images. It illuminates the strategy of photographic practices by which many tourists often respond to surrounding sites and sights. A camera is a universal instrument that results in seeing and making sense of the tourists' experience. Tourists do not want to take the risk of missing the sights/sites that the guidebooks inform them they need to 'see'. Some study tourists claimed that they checked the postcard racks to check out the 'Englishness' on display and then use their cameras to duplicate the shots (see Figures 11.1 and 11.2).

Figures 11.1 and 11.2 show that the tourist gaze is influenced by circulating flows of images and the belief in 'expert' photographic knowledge. In my own observations, the image of a place influences the numbers of photographs taken there. This image of a particular sight/site may come from various sources. including travel magazines, tourism brochures, and postcards. The camera is the tool for tourists to appropriate the collective gaze so they do not neel left out in

Figure 11.1   Checking Englishness from the postcard racks

Figure 11.2   Englishness displayed on the postcards

their travelling experiences. It is the machine through which the self and Other may be defined, recorded and made intelligible.

Photography and Embodiment of Sociability

With the understanding that cameras hold multiple functions in relation to tourist performative acts, I now focus on the complex connections between photography and social interaction. The usage of photographs in tourist narratives provides insights into how social and cultural factors interact with tourist performance. Photography is just one mediation; the significant focus is on the social interaction. At the same time, knowing what people are talking about when they show or look at personal travel photographs, we need to ask: what kind of cultural representations are being produced within photographs? How can these photographs be used to revive and interpret tourists' experiences? Why are some photographs significant to tourists?

We are familiar with the sight of tourists with their cameras hanging around their necks, especially among groups of Asian tourists (see many photos in Parr 1995). Based on observation and group discussions, all the study tourists travel with their cameras and I was surprised at how many photographs they took during their four-week tour of Britain.[2] The study tourists say that they enjoy taking pictures so that they can share their touristic experiences with their families and friends when they return home.

In email exchanges among the 12 study tourists in a post-tour focus group, the use of holiday photographs as a social tool to connect with others is given as the most common use of photos. In short, they use photographs as an essential method to cultivate and maintain contacts with their tour group members and also with other international friends, such as host families and their language classmates. Sending their photographs to friends was the most common usage of their photographs. Some of them made their photographs into greeting cards or calendars for New Year gifts to give and send away. Photographs convince and inform, and they are self-created souvenirs, as Kevin Meethan writes:

> A photographic record provides the material witness so to speak, that one has seen the sights, been there and done that. In this sense, they are goods to be worked at beyond the immediate point of commodification, functioning in much the same way as other forms of souvenirs and mementoes. (Meethan 2001: 84-85)

---

2   In the case of the study tour that I led in 1999, they averaged 8 rolls of 36 exposure film. Among 30 study tourists, it ranged from 8 study tourists who took 15 rolls and at the other extreme 2 who used only 4 rolls of films. In the digital age with the popularity of digital camera, films seem to be replaced by memory cards.

Figure 11.3   Study tourist punting on the River Cam

All the study tourists recall that when they took pictures, they had their families in mind. Some tourists took certain snapshots because they were asked for by their families. Cindy, a 17-year-old high school student, showed me a picture of herself taken when she was punting on the River Cam in Cambridge at her sister's request (see Figure 11.3).

In much the same way as the act of shopping which I discuss elsewhere (Yeh 2003), many of the photos were for their families who paid for the tour. Through fieldwork and numerous conversations with study tourists, I learned that most of the study tourists feel a strong obligation to take pictures for their families and friends. Pete, a 17-year-old high school student, says: 'My parents paid the tour for me so I want to give them something in return. Photographs are good gifts to share with them'. Grace said in the post-tour interview in 2001:

> I hate to see so many pictures of myself, but I had to do that for my family and my friends. My mum seems only to be interested in pictures with me in them, the rest of the pictures don't mean anything to her. She looked through the pictures quickly and I doubt that I really saw anything but me. (24 August 2001)

Adam, a 20-year-old university student, shares the same obligation:

> Basically, I don't want to take pictures because the postcards are good enough. But I have to take pictures for my family, especially my mum and my younger sister. They like Cambridge because of that soap opera.[3] But I think I also want to show them to my friends. With photos, it's easier for them to understand what I'm talking about. They also prove I've been in Cambridge. Joyce, can we take a picture together? (29 July 2000)

Antun, a 23-year-old university student, claims that taking pictures serves a special purpose in preparing his sister to come to Britain. He explains:

> Some of the pictures that I took are for my younger sister, especially those pictures at Cambridge. She's planning to come over for a visit next year. My pictures can give her some guidelines, though she's already familiar with all the views. She has been collecting all the pictures and posters. (2 August 2000)

This sense of obligation of taking pictures for others may lead to feelings of both pleasure and anxiety. Taking pictures, like collecting souvenirs, which I addressed in my doctoral thesis (Yeh 2003), is one of the study tourists' obligations to those who are unable to participate in the tour. These Taiwanese tourists brought plenty of blank rolls of film with them to Britain so that they would be able to record their experiences. The price of a roll of film in Britain is almost 2-3 times more than in Taiwan. Alice, a 21-year-old university student, regrets not bringing enough films from Taiwan to fulfil her goals of taking pictures and she said, 'I only brought 6 rolls of film and I'm so regretful now. I took about 10 pictures a day. But if we go sightseeing I take more'. In all the above examples, there is a fundamental link between the self-evidence of being there and the responsibility of showing images to others. The strong bond between family and the individual in Asian societies provides an explanation for the obligation of Asian tourists to take pictures for family and friends. Although this tourist performance might be more diverse in other types of tourists and societies, the cultural influence of these photographic practices is obvious.

Of the 110 photographs tourists chose to represent their sense of Englishness, 82 are of people. Images relating to study tourists themselves are especially important, such as self-portraits, group shots, host families and language teachers and new friends they made in England. Most of the pictures that I have seen are of people, including self-portraits before buildings, sites and sights (see Figures 11.4 and 11.5).

---

3   The soap opera is based on Chang's memoirs: *Bound Feet and Western Dress: A Memoir*, in which some of the setting is in Cambridge.

Figure 11.4   A group of Taiwanese study tourists and English building

The photographic images that the tourists showed me suggest the emphasis is on the significance of 'interaction' with others. Sally, a 20-year-old university student said:

> I took a lot of pictures of myself and 'foreigners'. [Here 'foreigners' mean any non-Taiwanese] They are cute and some are very stylish. Diana and I sometimes approach them and ask if I can take pictures with them even though we don't know them. (21 July 2000)

Indeed, people, including study tourists themselves, have become fetishized to represent study tourists' experiences of Englishness, which also reveals how emotional sense is closely connected to the 'tourist gaze'. These chosen 'iconic' images of people shape study tourists' sense of Taiwanese Englishness. None of the people they chose are from the British Royal Family, celebrities or political figures in Britain, but common and ordinary people going about their everyday lives.

It is these 'real' people whom they travelled with or encountered in England that shape their English cultural 'landscape' and how they 'view' Englishness. It seems to demonstrate that what is more important to Taiwanese study tourists is not only physically being in England, but also the 'authenticity' of their emotional involvement with England. There are 28 out of 82 'people' pictures of the host family, and indeed they are the study tourists' first contact of Englishness and play an important role in constructing their experiences of living in England. Group

Figure 11.5   Display of Englishness?

shots with the group members and new friends are also a top choice – 24 out of 82. There are 16 photographs of language teachers, even though some of them are American or Canadian. There are 14 pictures of study tourists individually at famous attractions, sites or in their rooms in the homes of the host families. The images of Taiwanese Englishness focusing on such encounters with people demonstrates the need to feel attached to others particularly in a new country. Emotions of belonging and welcoming play significant roles in shaping the tourist 'gaze' and experiences of Englishness.

The stereotypical image of Asian tourists is of them taking pictures of themselves on their holidays abroad (see Parr 1995). This characteristic also fits the young Taiwanese tourists. Undoubtedly, however, there is a cultural difference in their decision to photograph themselves or not. The desire to be photographed illustrates the differences between various cultures. Asian symbolic attitudes toward the family relationship make picture taking within the specific context of experience meaningful to the whole family. Study tourists, as we have noted, use the camera as a tool for social interaction and the products, photographs, serve the same function for Taiwanese study tourists. It is for the same reason that study tourists put in extra effort to take many pictures when they are abroad.

The chance to travel abroad has made study tourists more photography-conscious and interested in collecting specific sights/sites among themselves by using their cameras.

Holiday snapshots are never just souvenirs. We need to consider their levels of productions, interpretations and representations of tourist's experiences. Through photographic practices, Taiwanese study tourists make their own constructed images of English culture: self-referential systems of representation of Englishness which they take home and display to multiple audiences. Holiday photographs act as indexical signs of self-expressions that frame the tourist gaze, and thus contribute to how others understand tourists' experiences in various settings. Recognizing how holiday photographs are used by individuals to form satisfying relationships and convey information about Taiwanese Englishness, I suggest that tourists take an active role in creating 'social capital' (Bourdieu 1984) which is both a mark of personal distinction as well as a way of framing, interpreting and representing their socio-cultural encounters and networks.

For those who do not travel, but see the travel photographs, the photographs are an opportunity to 'dream into it, causing it to become subjectivized by the viewer's desires, memories and associations' (Osborne 2000: 77). Every photograph is in a sense a document of one segment of a tourist's life. They record the study tourists' emotions and such experiences as the houses where they stayed and what they ate with their host families. Therefore, photographs provide others with a frame of reference for understanding the travelling stories. Photographs can enact story-telling and can also offer a vivid sense of 'being there' for both story tellers and listeners. In this case, travelling stories are also important, particularly to those who do not travel, as Trinh writes: 'Tale-telling brings the impossible within reach' (1994: 11). Trinh summarizes:

> Travellers' tales bring the over-there home, and the over-here abroad. They not only bring the far away within reach, but also contribute…to challenging the home and abroad/dwelling and travelling dichotomy within specific actualities. At best, they speak to the problem of the impossibility of packaging a culture, or of defining an authentic cultural identity. (Trinh 1994: 22)

In this sense, travel stories and photographs provide non-travellers with windows on the world for 'encountering' people and 'experiencing' other cultures. Travel and photographs are, therefore, conceived here as being the process of human communication and social interaction.

For those who travelled together, an arranged meeting to exchange pictures after the tour suggests conversational gambits and allows more active personal relationships to develop among the tour members. In the following case Sylvia stated in our e-mail correspondence that she and other tour participants developed their friendship after sharing the photographs with each other. She wrote:

It was really good. Everything was about the study tour and we were very happy to see each other again...David [another study-tourist] took so many similar pictures to mine. Like old friends, we joked about each other. Pictures brought us closer...We were in Cambridge together and we knew the stories behind those pictures. It was different from showing the pictures to my family or classmates. (email correspondence on 10 February 2002)

Thus, the act of photographing involves another complex cultural reasoning: that of constructing personal relationships. Photography can be regarded as a social practice, as photographs mediate social relationships. I now turn to explore in what ways tourist photographs are a source of connection with others as well as with tourists' everyday life when they return home after their tours.

Mary Price suggests 'any photograph invites some comment' (1994: 45) and therefore, 'describing is necessary' (1994: 5). She argues there is no definitive meaning for a photograph. These fragmented life moments or travel events demand description, interpretations and response. And yet, the meanings and comments on a photograph also involve how viewers 'read' or 'look' at it and incorporate their point of view. Photographs provide conversational topics to fill in the gaps between the past 'then and there' and the present 'here and now'. The work of Annette Kuhn (1995) shows how private family photographs evoke public cultural memories and generate discourses of meaning-making. To her, a photograph 'sets the scene for recollection' (1995: 18). Despite the fact that each person articulates his/her photographs in a particular manner, this shows that they are not strictly individual, but also collective. In the same manner, study tourists use their travel photographs to initiate communication and reconstruct memories. Anna, a 21-year-old university student, says:

I put some pictures on the walls in the dorm. Every time there's a visitor in our room we start talking about my experiences of the study tour because they see the pictures. I've made many friends through talking about my pictures and the summer that I spent in Britain. (24 August 2001)

Holiday photographs are also a powerful medium of memories. Many of the study tourists I interviewed use their travel photographs as resources to bring their memories back. Rather than a source of nostalgia, photographs are methods for study tourists to lift their spirits. Alice says:

I don't know why. But every time I look at my photo albums of the study tour in England I feel 'high I'm really happy to look through them. I had a great time there and all the happy memories come back when I flip through each page of the albums. Even now I'm talking to you I can feel the happy feeling come back to me. (9 September 2001)

Some study tourists share this happiness with their friends. When I conducted a post-tour interview with John who is doing his military service, he told me:

> Those pictures that I took in Britain are great 'saviours' in the boring army life. I talked about the stories of the pictures again and again to entertain myself and other pals. By doing that I feel I'm entering another interesting space and different time zone which makes my tough but boring army life more endurable. My photographs are helpful for my pals too. They are happy to look at them ane listen to my stories. Some of them also plan to visit Britain 9after we finish our military service) because of my photographs and stories. (1 September 2001)

Photography fragments time and it fractures space as well. When a tourist takes a picture, he/she freezes the moment of time and segments the space or a particular landscape. The act of taking a photograph is always at the present moment, but the 'now' always becomes 'then'. And yet, 'then' can be traced and linked into a continuous 'present' when travel photographs are displayed. Travel photographs exhibit the past experiences of travelling to the present moment and expand the experience as long as tourists wish. Looking at photographs, as John Taylor suggests, 'invites imaginative travel into more or less distant pasts' (1994: 4). Such imaginative travel is also associated with 'freedom' and 'escape'. By doing so, the moment of the past is performed and is connected to the present which gives travelling experiences a sense of continuity of 'freedom' and 'escape', conventionally associated with touristic practices.

Conclusion

I have sketched out how cameras and photographic discourses may construct social interaction. Tourists' travel photographs are significant in constructing social relationships and the photographic tool, the act and productions, can be seen as tourist performance and experiences of collecting Englishness. I consider that photography is not only a means of visual consumption, but also an embodied practice of sociability, which reflects certain specific cultural values and practices. The use of the tourists' own holiday snapshots as research material and the opportunities to view study tourists' travelling photo albums to provide such a context does not imply that the representation is a definitive interpretation of Englishness; it simply means that this context is a lens, a filter through which we can view how travelling experiences provide social interaction for particular tourists. By examining the various functions of a camera I have identified photographing as one of the tourist's performing acts and have demonstrated ways in which tourism and photography have come to influence each other. I have also investigated how Taiwanese tourists use the camera as a means of picturing social relations, and employ the act of photographing to construct their travel performances and to mediate their visual encounters with Englishness-as-Otherness. Indeed, the camera is used by tourists extensively and is an important contributor to tourists' visualizing of Otherness and a vehicle to represent what they see. For tourists, cameras can also be used as tools for examining cultural differences and bringing a scene alive. They can record familiar or 'extraordinary' objects, people, the land/ townscapes, architecture and interiors of houses where they stay, and they can be a means of informing and envisioning. The photographic acts that Taiwanese study tourists performed have been examined to highlight the Taiwanese tourist gaze upon Englishness.

As a performative practice, cameras have symbolic associations and signal social relationships. Cameras are tools, signifiers and communicators. They speak for identities and construct the tourists' images of being temporary visitors. With cameras, study-tourists are easily recognizable as strangers who do not belong to the place they visit, but cameras also create group ties. On the one hand, cameras are a vehicle for tourists to perform differently from the locals and also to demonstrate that they are physically embodied in different cultural settings. On the other hand, cameras can mediate personal and social relations. Cameras can also function as symbolic expressions of tourists' individual tastes, with photographs as documents which can be evaluated as the products of creativity and the judgment of taste. The types of cameras and even the numbers of cameras that tourists travel with can be used as codes to distinguish different types of tourists, a matter which needs to be addressed and subjected to more empirical research.

Bibliography

Albers, P. and James, W. (1988), 'Travel Photography: A Methodological Approach', *Annals of Tourism Research*, Vol. 15, pp. 134-158.
Bourdieu. P. (1984), *Distinction: A Social Critique of the Judgement of Taste*, Routledge, London.
Bourdieu. P. (1990a), *The Logic of Practices*, Polity Press, Cambridge.
Bourdieu. P. (1990b), *Photography: A Middlebrow Art*, Polity Press, London.
Chalfen, R. (1979), 'Photography's Role in Tourism: Some Unexplored Relationships', *Annals of Tourism Research*, Vol. 6, pp. 435-447.
Cohen, E., Nir, Y., and Almagor U. (1992), 'Stranger-Local Interaction in Photography', *Annals of Tourism Research*, Vol. 19, pp. 213-133.
Foucault, M. (1973), *The Order of Things*, translated by A. Sheridan, Vintage, New York.
Foucault, M. (1977), *Discipline and Punish: The Birth of the Prison*, Penguin, London.
Horne, D. (1986), *The Public Culture: The Triumph of Industrialism*, Pluto Press, London.
Kuhn, A. (1995), *Family Secrets: Acts of Memory and Imagination*, Verso, London.

MacCannell, D. (1973), 'Staged Authenticity: The Arrangement of Social Space in Tourist Setting', *American Journal of Sociology*, Vol. 79, pp. 589-603.

MacCannell, D. (1976), *The Tourist: A New Theory of the Leisure Class*, University of California Press, Berkeley.

Markwell, K. (1997), 'Dimensions of Photography on a Nature-Based Tour', *Annals of Tourism Research*, Vol. 24, No 1, pp. 131-155.

Meethan, K. (2001), *Tourism in Global Society: Place, Culture, Consumption*, Palgrave, New York.

Osborne, P. (2000), *Travelling Light: Photography, Travel and Visual Culture*, Manchester University Press, Manchester and New York.

Parr, M. (1995), *Small Worlds*, Magnum, London.

Price, M. (1994), *The Photography: A Strange, Confined Space*, Stanford University Press, Stanford.

Sontag, S. (1979), *On Photograph*, Penguin, Harmondsworth.

Spence, J. and Holland, P. (eds) (1991), *Family Snaps: The Meanings of Domestic Photography*, Virgo, London.

Taylor, J. (1994), *A Dream of England: Landscape Photography and Tourist Imagination*, Manchester University Press, Manchester.

Trinh, Minh-ha T. (1994), 'Other than Myself/My Other Self', in G. Robertson, M. Mash, L. Tickner, J. Bird, B. Curtis and T. Putnam, (eds), *"Travellers' Tales"*: *Narratives of Home and Displacement*, Routledge, London.

Urry, J. (1990a), *The Tourist Gaze: Leisure and Travel in Contemporary Societies*, Sage, London.

Urry, J. (1990b), 'The 'Consumption' of Tourism', *Sociology*, Vol. 24, pp. 23-35.

Urry, J. (1995), *Consuming Places*, Routledge, London and New York.

Urry, J. (2002), *The Tourist Gaze, 2nd edn*, Sage, London.

Yeh, J.H. (2003), *Journeys to the West: Travelling, Learning & Consuming Englishness*, PhD Thesis, Department of Sociology, Lancaster University.

Yeh, J. H. (2009), 'Still Vision and Mobile Youth: Tourist Photos, Travel Narratives and Taiwanese Modernity', in T. Winter, P. Teo and T.C. Cheung (eds) *Asia on Tour: Exploring the Rise of Asian Tourism*, Routledge, London, pp. 302-314.

Chapter 12

Disposable Camera Snapshots: Interviewing Tourists in the Field

Elisabeth Brandin

Introduction

The purpose of this chapter is to discuss practical as well as theoretical implications of a fieldwork method for collecting information from tourists. The idea was to interview canoeing tourists immediately following their canoe trip, using their own snapshots as a basis for the conversation. To that end, disposable cameras were handed out to canoeing tourists prior to their trip, and films were developed as soon as their trip ended. The challenges prompting this method grew out of two intertwined sets of circumstances. The first pertain to the research questions of the study. The objective was to explore variations in canoeing practices and to explore variations in meaning of these practices. The second pertain to the place's infrastructure, the infrastructure of canoe tourism in the chosen area, and who the tourists were, all of which informed the choice of how to collect information from tourists. The information collected from interviews based on tourists' snapshots in the present study is theoretically based on ideas of embodied practices and their semiotics. The methodological implication is that the photographs and interviewees become inseparable. Within such a frame, the results relate to the importance of using the tourists' own photographs as a starting point for the interviews: the interpretation of photographs cannot be wholly separated from the photographers' interpretations, the combination of interviews and photographs can elicit information that goes beyond what appears in the photographs, some control over the situation is shared by the interviewees, cameras are less intrusive than an ethnographic researcher would be, and constraints of time and language barriers are negotiated.

This chapter deals with the use of tourists' own photographs in a field interview situation for collecting empirical information, and with situating this approach methodologically. The method of analysis of the empirical data lies outside the scope of this chapter. In the following I will first discuss the theoretical challenges related to doing interviews with tourists in a field situation based on photographs taken by the tourists. I discuss seeing practices in relation to embodied practices and how tourists negotiate extraordinary experiences by taking 'ordinary' photographs. Next, I discuss some practical challenges and how these influenced the choice of method. Following that, I describe and comment on the practicalities of interviewing in the field. Finally, the chapter closes with some concluding reflections. The method for collecting empirical material was part of a study with the objectives of analysing signifying practices of several actors, including canoeing tourists, and exploring how canoe tourism becomes spatialised (Brandin, 2003).

Between Seeing Practices, Embodied Experiences and Verbal Interviewing

It may seem contradictory to use a visual method as part of collecting data within a theoretic frame of embodied practice. By putting a lens between oneself and the environment or an event, seeing practices become dissociated, as in the tourist gaze (Urry, 1990). Events and environments are reduced to what can be seen in the frame, and experiences may be missed because of the preoccupation with taking pictures (Chalfen, 1987; Urry, 1990). Seeing practices have further been explored as the short-term, fleeting seeing during motion, the tourist glance (Larsen, 2001), as oppressive (male) looking, and as voyeurism (e.g. Hall, 1997b). Embodied practice is, on the other hand, associated with immediacy and intimacy, with everyday life, and personal experiences using multiple senses (Crouch, 1999; Crouch, 2000; Macnaghten and Urry, 2001; Thrift, 1996: 1-50). Within this frame seeing can be thought of as engaged seeing, where the viewer is part of her/his surroundings (Chaney, 2002). Crang (1997) explores tourists' photographic practices as engaging with learning spatially through embodied practices. Seeing practices can be part of both representational theorising and non-representational theorising. While seeing is certainly embodied, embodied practices include other senses and other ways of knowing than those associated with textual practices. The focus on gaining knowledge through visual and oral practices in this study relies on knowledge being communicated through signifying systems (Hall, 1997a). Gaining knowledge about embodied practices by visual and oral signifying systems alone may provide only partial knowledge. It is maintained that knowing through the body also involves knowing through the senses in a way that cannot be communicated through signifying systems, as such knowledge is non-representational (Crang, 1999; Crouch, 2000; Thrift, 1996: 1-50).

Tourism is theorised in terms of seeing the extraordinary (Urry, 1990). Tourists' pictures may, on the other hand, often be 'ordinary' in the sense that tourists often take photographs of the same sights, monuments and events. The meaning of the extraordinary becomes personal. The photographs become important only to close friends and relatives, and they act as documentation of the successful life (Chalfen, 1987). Andersson Cederholm (1999) shows how backpackers personalise their 'ordinary' photographs of tourist sites. Practices of exoticising make it possible to negotiate the contradiction of ordinary and extraordinary to fit into the discourse of backpackers' successful trips. Pocock (1982) finds that people tend to remember views as more colourful, impressive and three-dimensional than they appear in photographs. Photographs trigger the memory and provide proof that the experience occurred, but they are no substitute for the experience (Chalfen, 1987). Although photographs can never replace an experience, they work not only metaphorically but also, in a way, metonymically. The small two-dimensional photographic image stands for the three-dimensional and multi-sensory lived experience. As such, photographs can spark memories of the lived experience related not only to what is depicted within them, but also to what is not shown in them (Andersson Cederholm, 1999; Chalfen, 1987).

Using photographs as a method of collecting information by showing them to interviewees for responses, where the photographs are not taken by the interviewees, is a known strategy (Collier and Collier, 1986). This method is also used in tourism research (Dann, 1995; Crawshaw and Urry, 1997; Fenton et al., 1998). Interviewees' own photographs have been used to answer questions on identity (Ziller, 1990; Clancy and Dollinger, 1993; Dollinger et al., 1996; Dollinger et al., 1999) and to explore daily transactions in a children's environment (Aitken and Wingate, 1993). Chalfen (1987) explored peoples' lives, including their experiences as tourists, by including interviews based on photo albums. Further, people's own photographs have been used for empowerment, increasing collective knowledge and informing policymakers with the creation of so called photo novellas (Wang and Burris, 1994; Wang et al., 1996). Interviews over tourists' own photographs have been done post-trip at home (Chalfen, 1987; Markwell, 1997) and in field situations on trains as well as post-trip home situations (Andersson Cederholm, 1999).

While the negative aspects of photography and the limits of the photograph in relation to tourists' practices and experiences cannot be neglected, I want to call to attention the possibilities of tourists' own photographs. The possibilities as I find them relate to the research question dealing with variation and with detail. The objective of the study was to gain knowledge not only about variations in what canoeing tourists did, but further to explore some of the variations of meanings of what tourists did, and subsequently to analyse these varying meanings for possible common themes. When studying variations, not only extreme practices but all kinds of practices of the relative 'everyday' aspects of canoeing leading to tourists' experiences through multiple senses were of interest to me, hence the study of embodied practices. To gain detailed information about the variations of meaning attributed to what the tourists did, the information was based on the tourists' personal experiences. To collect sufficiently detailed information, some kind of in-depth interviewing or ethnography would be possible alternatives. Using tourists' own photographs as the basis for interviews was a way to gain details as well as to bring about discussion of experiences that went beyond those shown by the photographs. I feared that information gathered from interviews only, without the use of the tourists' photographs, might be less detailed. The photographs prompted memories of other experiences not captured in the pictures. Ethnography was not an option mainly because of the tourist infrastructure in the area, which leads to a discussion of the practical challenges related to doing conversational interviews with tourists' photographs as the starting point.

The Challenges of the Place and the Tourists

The canoeing area in question, Dalsland-Nordmarken, is easy to get to. Situated in south-west Sweden, it is close to the continent and connected by the Öresund bridge to Denmark as well as by ferries to Denmark and Germany. Many roads cross the 6000 square kilometres of land and water. At its heart are the connected lakes of the Dalsland Canal and their adjoining lakes in a forested rift-valley landscape. The lakes are connected by 31 locks and one aqueduct, only 10 kilometres are dug. An area of approximately 600 square kilometres of water is used for canoeing. There is an array of protective measures with legislative support, but no comprehensive legislation or regulation as it is not a national park or a reserve. There are various regulations and codes of conduct with varying types of support in legislation. The area is accessible for canoeing due to the Swedish right of public access, which grants individuals access to nature.

The area is popular among international visitors, who come here to canoe. In fact, 85 per cent of the visitors are international. Most come from Germany and Denmark. About 25,000 canoeists come to the area every year, staying an average of 3 to 4 days each. The first canoeists of the season arrive in May, and the last leave in September. The season picks up after midsummer and peaks in July. Canoeing tourists vary not only in nationality, but in other ways as well. They vary with respect to the amount of time they spend out paddling. The group size varies. Some canoe with friends and relatives, others with groups of people they don't know. Some book trips in advance, others don't. Some groups are affiliated with churches, some are scouts or school classes. Some go on low-cost trips with non-commercial groups, others travel with commercial tour organisers. There are experienced canoeists and beginners, people who have been in the area before and people who haven't. In 1999, the 18 canoe rental firms based in the area had a total of 1300 canoes available for rent, with individual firms renting out between 12 and 200 canoes. It is assumed that some 600 to 700 additional canoes are brought into the area by tour operators who are not based in the area, by organisations and by individuals. To alleviate crowding and to protect the environment, a hundred campsites have been built over the years, each able to accommodate ten tents. These sites provide log shelters, approved sites for campfires, modern dry toilets and logs for firewood. Further, rangers are employed by the municipalities for the benefit of the tourists. By law, access to nature cannot be charged for in general. A system of charging for campsite facilities and services covers about a third of the total management costs, including compensation for landowners.

The easy access to the area, the proximity to the continent and the suitability for easy paddling together with marketing efforts have attracted many international tourists. The choice of doing conversational interviews based on tourists' photographs grew out of the practical challenges associated with doing in-depth interviews covering such abstract topics as personal meanings of practices with the predominantly international tourists to the area. Further, a meaningful interview takes time, and time may be limited for tourists, who may have many activities planned during their stay. As tourists were interviewed after they completed their canoe trip, ferry timetables posed limitations to interviews. Although the area can be crowded at times in some places, it was a challenge to, within a few weeks, select participants from a wide range of circumstances who also had enough time to participate in a post-trip interview before travelling on. The idea was that a relatively meaningful conversational interview could be struck up in a shorter time using the tourists' own photographs, which could act as memory aids and also contribute to a more relaxed situation where trust between interviewer and interviewees could be fostered (Andersson Cederholm, 1999).

Interviewing using photographs was favoured over ethnography on several grounds. To follow specific groups as they paddled would have been a method for gathering detailed information because it would allow the researcher to spend a longer time with the tourists. This method was rejected on the grounds that it would be too time-consuming and it would have been intrusive to accompany small groups of friends, families and couples on vacation (although it may not have been a problem with larger groups). The option of making observations and doing interviews while staying at one or a few campsites, waiting for tourists to come, was dismissed because canoeists tend to avoid campsites that are already occupied and try to find their own sites, unless there are many people in the area. Moreover, this method would have excluded those who prefer to avoid established campsites, and this group's ideas of what canoeing is about may differ from the ideas of those using such campsites. Using a boat to seek out tourists to observe and talk to canoeists camping in a variety of sites was also dismissed as the area is large and it would have been too time-consuming. Another option would have been to contact tourists after they returned home and conduct in-depth interviews, but while such an approach would have been possible, it would also have been laborious and expensive, and would have involved finding someone to do interviews in the native languages of the international tourists. If this approach had been adopted, the tourists' snapshots taken with their own cameras could have been used as a basis for conversation.

Instead, a middle way was sought. The practical use of disposable cameras, processing film in an hour and doing field interviews with international tourists based on their snapshots is the topic of the next section.

Interviewing in the Field

I provided canoeing tourists with disposable cameras, which I instructed them to use as if they were their own. I explained that their photographs would be used as the basis for interviews with them about what they had done and experienced during their trip. For eight of the high-season weeks for canoeing in 1999, I distributed 13 single-use Kodak Fun Flash cameras (27 exposures) to canoeing tourists before they set out on their trip. The cameras were collected on the tourists' return, and the films were developed with double prints within an hour.

Semi-structured conversational interviews were then conducted about the canoe tour just completed, using the snapshots as a basis for the interviews. The tourists were allowed to keep the negatives and one set of prints (at the time, the cost was 23 EUR), while I kept one set of prints for the study. All the cameras were returned and all participants completed the interview. As part of the interview, the tourists were asked to choose the ten photographs from the developed film that they thought best captured what they had done and experienced. The interviews took between one and two hours. All the interviews were taped (bar one, where there was a technical problem with the recording equipment) and subsequently transcribed for later analysis.

Pre-interview Concerns

Using disposable cameras meant that a relatively large number of cameras could be distributed over a short time at a limited cost. By distributing disposable cameras and instructing the tourists to use them as if they were their own, I could get information from a relatively large number of tourists in varying circumstances at a limited cost without being present myself to record it. The cameras worked as my less intrusive 'deputies'. I could distribute cameras to more tourists than I could myself have followed, and I could monitor and secure variation by choosing who would get cameras.

Most of the interviews were conducted only an hour and a half to two hours after the canoeists had finished their trip. This could be done because of the availability of quick, relatively inexpensive print development, although business hours affected who could be interviewed. The tourists had to set aside a minimum of three hours for the interview, the time for processing the film and the time needed to travel (by car) to and from the two available photo shops in the area. Further, the offers on double prints at the time of developing the films made it possible to offer an immediate reward to the tourists and aided the efficiency of the study.

The study depended on the technologies that made available inexpensive, easy to handle cameras able to produce pictures of acceptable quality, as well as fast (within an hour), affordable film development. The cameras were readily accepted by those who did not bring their own cameras. Tourists who had brought their own, more sophisticated, cameras were asked to use only the disposable camera I provided. Some tourists reported that they had complied with that request, others reported that they had not. Although the cameras were equipped with flashes, they were reported not to operate properly, and range for the best photo quality was not always followed, resulting in greyish pictures. Cameras were not waterproof, and so they were often packed away while the tourists were actually canoeing or when it was raining.

One drawback of instructing tourists to use the cameras as if they were their own, instead of instructing them to take pictures of anything important to them, is that important experiences may not have been caught on film. Important activities and experiences become conflated with taking the pictures. On the other hand, asking tourists to take pictures of anything they feel is important may have presented some difficulties in practice as there is an element of the unexpected in tourism. Pictures cannot always be planned, and cameras may have been packed to protect them from the water, so some photo opportunities could easily be missed. Nevertheless, the tourists had some latitude with regard to taking the pictures. The procedure implies that to some extent control over the research situation was transferred from the researcher to the researched. The usual researcher (subject) – researched (object) relation thus becomes less distinct. This position may have left tourists compelled to include or leave out certain motifs.

Concerns During the Interview

Once the photographs had been developed and before the actual interview began, I asked the tourists to choose the ten photographs they thought best described what they had done and experienced during the trip. This was done in order to limit the length of the interview and to focus the interview on what had been perceived as most important. Then the interview commenced, and the tourists chose the order in which the photographs would be discussed. The sorting of the pictures and the decision about the order in which to talk about them transferred some control over the situation to the interviewees (Andersson Cederholm, 1999). Sorting the pictures opened up the possibility of discussing why certain photographs were not chosen for discussion. I had little time to pursue this line of questioning, but when I did it enriched my understanding of the meaning of the experiences. This is a strategy I believe is well worth exploring further.

Ziller (1990) took the approach of specifically limiting the number of pictures taken by having subjects select motifs before the photographs were taken. The number of photographs required were three, six and twelve, answering questions such as 'Who are you?', 'What does the good life mean to you?', 'What does woman mean?', 'What does war mean?' and 'Me and my world'. The photographers were then asked to comment on the photographs at varying length. Here the discrepancy between the research question and the photographs is limited by the photographers' conscious decisions about what to include and what to exclude. My approach is less direct and in some ways deprives the photographers of being consciously part of and controlling the research questions. My reason for not being more explicit was in part the exploratory nature of the study and in part respect for the tourists, as tourists would not want to be burdened with what could have been perceived as a task that would intrude on their canoeing experience. Further, to ask tourists to plan to take certain types of pictures might have proven difficult, since tourism is to some extent about the unexpected. Since, by definition, the unexpected cannot be anticipated, the tourists may not have been prepared to capture certain things on film. The tourists did, however, share some control over the interview situation by selecting the developed photographs that would provide a basis for discussion.

My questions were not geared only towards the pictures, but also included research questions geared toward identifying variations. The conversations took off in several directions, depending on what the tourists talked about and specific questions I asked. This strategy resembles that of Andersson Cederholm (1999) but is different from that of Ziller (1990), who kept questions close to the pictures, the pictures being allowed their own non-verbal discourse as well as serving to delimit the study. The looser relation between photographs and practices in the canoe study allowed for questions about whether there were practices or experiences that for some reason were not included in the pictures, and why (leaving aside the technical problems discussed above). Some situations can be very private, and a sense of not wanting to destroy or disturb something might inhibit the taking of pictures. I asked the interviewees if there were any important moments they did not capture on film and, if so, whether they would like to talk about them. Of course, there is always the possibility that interviewees might not be prepared to disclose this information to the interviewer. However, most interviewees answered that no important experiences were omitted. One interviewee, asked why she did not have any photographs without people, said that the pictures of nature were in her head and that no photograph could compare with her mental picture, and so she avoided this type of photograph (Pocock, 1982). Such statements led to further discussion about nature experiences.

Interviewees may have felt gratitude because they benefited from the arrangement. To try and find out whether those interviewed did think about who paid for the camera and the purpose for which it was handed out, the interview included questions on whether they had specifically taken some pictures for me or specifically refrained from taking others because of me. Most reported not thinking about whose camera it was. One person showed me a picture that he said was specifically taken for me, because it was an experience he had had before and wanted to show me because it was important to him. Otherwise he would not have taken the photograph as he already had similar photographs. Another person specifically apologised for the content of some pictures, saying that he had forgotten that the camera was not his own. One interviewee said that he had been careful to spread out the photographs evenly over time.

The photograph's ability to trigger memories is intertwined with the time of the interview. Developing the films within an hour made it possible to conduct interviews close to the end of the tour. As time passes, the meaning of practices and experiences may change. Pocock (1982) saw changes in the remembered size of objects and the intensity of experiences. Andersson Cederholm (1999) discusses changes in backpackers' narratives of experiences over time as the narratives progressively conform to 'backpacker culture'. My interest was in the variety of practices and experiences, both general and specific, rather than in changes in meaning or in importance. Had the interviews been conducted at any other time, meanings might have changed, although the story produced would have been no less true. If interviews at a later stage had produced fewer details, research findings would have been affected. To the extent that details are easier to remember when photographs are present, conducting the interviews later may profit relatively more from the use of photographs than interviews conducted closer to an event.

The tangibility of the photographs was important in several ways. The photographic prints and the maximum of 27 exposures facilitated doing the interviews under field conditions. For the most part the interviews were conducted outside, sitting on the grass or at a table. The number of prints and the format allowed us to spread them out and have an overview not only of the ten chosen photographs but also of the ones not chosen. Another important role, mentioned above, is the reassuring effect handling the pictures had on both interviewer and interviewees. The interviewees shared some control over the situation, which may have made it easier to develop trust. This was discussed above as having specific importance as many interviews were conducted within strict time constraints and in languages foreign to both the interviewees and myself. Three awkward situations developed due to communication difficulties. One interview was conducted in English and translated back and forth from French by a friend in the interviewee's canoeing party. In this case, having the photograph was reassuring for all three of us, as we could point to the pictures and ask questions. In another case, the interviewee was not able to elaborate about experiences and meanings in English, and in yet another, I felt unsure as to how well we had understood each other. This shows that interviewing based on photographs has its limits and that the information from the thirteen interviews varied. Despite the difficulties in these three interviews and the range of variation in the material of all thirteen interviews, every interview added something to the subsequent analysis.

Concluding Reflections

This chapter has revolved around a method of collecting information from tourists in a field situation where the focus has been on the inseparability of interviewee/ photographer and photograph. The research question concerned exploring detailed variations in meanings of canoeing tourists' varied practices during the relative 'everyday life' of a canoe trip. Theoretically, both the research question and the method used to gather information rest on the assumption that embodied practices of tourists can be communicated. Further, it is implied that this communication would be facilitated in several ways by doing conversational interviews based on tourists' own snapshots. It was expected that using the tourists' own photographs as a basis for interviews would prompt the tourists to recall detailed personal experiences, which would provide variety. Further, variation was thought to be gained by giving interviewees greater opportunities to relate personal experiences from the trip that go beyond what is shown in the photographs.

Practically, in the field, it was possible to obtain variation and overcome time constraints by distributing several disposable cameras that could be used by different tourist groups simultaneously. Furthermore, developing films within an hour of each trip's completion made it possible to do interviews right after the canoe trips. This made it easier to limit the time requirements and find a wider variety of participants than would be the case if we used ethnography or sought out tourists at home. Cameras are less intrusive than a researcher, acting as a less conspicuous 'deputy' in comparison to ethnography. Photographs worked as intermediaries during interviews conducted under time pressure and in languages foreign to both interviewer and interviewees.

Interviewees gained some control by choosing what events to take pictures of and by being given the opportunity to choose which photographs would be discussed during the interview. The latter additionally provided an opportunity to consider why certain photographs were chosen and others were not, to gain more detailed and varied information. The tangibility of the photographs was important to negotiating time constraints and communication difficulties. Developing the film with double prints made it possible to reward the interviewees immediately.

To do conversational interviews with tourists' own photographs as the starting point in close proximity to a tourist trip may prove difficult when time constraints and frame of mind do not allow tourists to participate in interviews. A marked constraint of this study was the distance to the photo shops and their closing hours. On the other hand, the fact that canoeing tourists, on their return, took some time to unpack and clean the canoe and get ready for land transport allowed for time to rush off and have the films developed. Further, many tourists chose to stay on for a day or two to use the camping facilities in the area. These tourists who were not pressed for time were exceptionally helpful and eager to do interviews. Before using this type of method, practical circumstances need to be thought through.

The method described relies on the intertwining of words and photographs. Chalfen (1987) notes that the presentation of photographs to an audience in a home situation is never a silent matter. In the discourse of how home photography is used, photographs are always accompanied by comments. Further, photographs may also be accompanied with texts such as captions or even written manuscripts for more formal talks. Similarly, tourist photography is not silent. The photographs are used as a basis for discussions for the specific purpose of strengthening social bonds and showing oneself and others that one's life is proceeding successfully (Andersson Cederholm, 1999; Chalfen, 1987; Markwell, 1997). Thus conducting conversational interviews in the field based on tourists' own photographs might not be an entirely strange situation to tourists. Photographs facilitate talk; they may even promote talk. As in the home situation where photographs are shown to friends and family, they trigger memories and the resulting discussion may widen.

References

Aitken, S. and Wingate, J. (1993), 'A preliminary study of the self-directed photography of middle-class, homeless, and mobility-impaired children', *Professional Geographer* 45(1), 65-72.

Andersson Cederholm, E. (1999), *Det extraordinäras lockelse: Luffarturistens bilder och upplevelser. [The Attraction of the Extraordinary: Images and Experiences among Backpacker Tourists]*. Lund: Arkiv.

Brandin, E. (2003) *Spatializing Canoe Tourism: Negotiating Practices in Dalsland-Nordmarken.* Karlstad University Studies 2003:8. Department of Geography and Tourism, Research Unit for Tourism and Leisure, Karlstad University, Karlstad.

Chalfen, R.M. (1979), 'Photography's role in tourism: some unexplored relationships', *Annals of Tourism Research* 6(4), 435-447.

Chalfen, R. (1987), *Snapshot Versions of Life*. Bowling Green, Ohio: Bowling Green State University Popular Press.

Chaney, D. (2002), 'The powers of metaphors in tourism theory', In: Coleman, S. and Crang, M. (eds) *Tourism: Between Place and Performance*, pp. 193-206. New York: Berghahn Books.

Clancy, S.M. and Dollinger, S.J. (1993), 'Photographic depictions of the self: gender and age-differences in social connectedness', *Sex Roles* 29(7-8), 477-495.

Collier, J. and Collier, M. (1986), *Visual Anthropology: Photography as a Research Method.* Rev. and expanded edn, Albuquerque: University of New Mexico Press.

Crang, M. (1997), 'Picturing Practices: Research through the Tourist Gaze', *Progress in Human Geography* 21(3), 359-373.

Crang, M. (1999), 'Knowing, Tourism and Practices of Vision', In: Crouch, D. (ed.). *Leisure/Tourism Geographies: Practices and Geographical Knowledge*, pp. 238-256, London and New York: Routledge.

Crawshaw, C. and Urry, J. (1997), 'Tourism and the Photographic Eye', In: Rojek, C. and Urry, J. (eds) *Touring Cultures: Transformations of Travel and Theory*, pp. 176-195, London and New York: Routledge.

Crouch, D. (1999) (ed.), *Leisure/Tourism Geographies*, London: Routledge.

Crouch, D. (2000), 'Places around us: Embodied lay geographies in leisure and tourism', *Leisure Studies* 19(2), 63-76.

Dann, G. (1995), 'A Socio-linguistic Approach Towards Changing Tourist Imagery', In: Butler, Richard and Pearce, D. (eds). *Change in Tourism: People, Place, Processes*, pp. 114-136, London and New York: Routledge.

Dollinger, S.J., Cook, C.A. and Robinson, N.M. (1999), 'Correlates of autophotographic individuality: therapy experience and loneliness', *Journal of Social and Clinical Psychology* 18(3), 325-340.

Dollinger, S.J., Preston, L.A., O'Brien, S.P. and DiLalla, D.L. (1996), 'Individuality and relatedness of the self: an autobiographic study', *Journal of Personality and Social Psychology* 71(6), 1268-1278.

Fenton, M.D., Young, M. and Johnson, V.Y. (1998), 'Re-representing the great barrier reef to tourists: implications for tourist experience and evaluation of coral reef environments', *Leisure Sciences* 20(3), 177-192.

Hall, S. (1997a), 'The work of representation', In: Hall, S. (ed.) *Representation: Cultural Representations and Signifying Practices*, pp. 13-74, London: Sage and Milton Keynes: The Open University.

Hall, S. (1997b), 'The spectacle of the other', In: Hall, S. (ed.) *Representation: Cultural Representations and Signifying Practices*, pp. 223-279, London: Sage and Milton Keynes: The Open University.

Larsen, J. (2001), 'Tourism mobilities and the travel glance: experiences of being on the move', *Scandinavian Journal of Hospitality and Tourism* 1(2), 80-98.

Macnaghten, P. and Urry, J. (2001) (eds), *Bodies of Nature*, London: Sage.

Markwell, K.W. (1997), 'Dimensions of photography in a nature-based tour', *Annals of Tourism Research* 24(1), 131-155.

Pocock, D.C.D. (1982), 'Valued Landscape in Memory: The View from Prebends' Bridge', *Transactions of the Institute of British Geographers* 7: 354-364.

Thrift, N.(1996), *Spatial Formations.* London: Sage.

Urry, J. (1990), *The Tourist Gaze: Leisure and Travel in Contemporary Societies*, London: Sage.

Wang, C. and Burris, M.A. (1994), 'Empowerment through photo novella: portraits of participation', *Health Education Quarterly*, 21(2), 171-186.

Wang, C., Burris, M.A. and Ping, X.Y. (1996), 'Chinese village women as visual anthropologists: a participatory approach to reaching policymakers', *Social Science and Medicine* 42(10), 1391-1400.

Ziller, R.C. (1990), *Photographing the Self: Methods for Observing Personal Orientations*, Newbury Park: Sage.

Chapter 13

Connecting Cultural Identity and Place through Tourist Photography: American Jewish Youth on a First Trip to Israel

Rebekah Sobel

Introduction

With strong financial support from the Israeli government and North American Jewish philanthropies, the Birthright Israel gift offers students free trips to visit Israel in order to foster their Jewish identity and connection to Israel. This trip was introduced in the fall of 1999 in an effort to encourage Jewish youth between 15 and 26 to identify with their Jewish heritage and become participants in the greater Jewish community upon returning from Israel. In June 2000, 19 Georgetown University students, along with almost 3,000 other college students from throughout the United States, attended a ten-day Birthright Israel trip organized by Hillel International. This study evaluates the travel experiences of those Georgetown University students through their sentiments, experiences and photographs. This analysis employs an innovative methodology in looking closely at students' photographs and experiences on the trip. Students were interviewed before, during and after their trip. Six months after returning from Israel, students were asked to describe their travel experience with their photographs. Considering students' photographs as both data and narrative context in describing travel experiences is central in exploring relationships between identity and place.

North American Jewish philanthropists Charles Bronfman and Michael Steinhardt, along with program and educational support from the umbrella organization United Jewish Communities, created Birthright Israel in order to influence 'existing "Jewish culture" by making an educational trip to Israel an integral part of every Jewish youth's life' (Taglit 1999: 3). Some concerns raised by the National Jewish Population Survey of 1990 (NJPS) inspired these founders to implement the program for Jewish youth in the Diaspora, aged 15-26. The survey revealed that intermarriage (between Jews and non-Jews) is increasing in America at a seemingly higher rate than previously reported in earlier census surveys (NJPS 1990). Results released in the 2000 survey argue the number of Jews in the US decreased 5% in the last 10 years. These founders are concerned that today's Jewish youth will continue to intermarry and assimilate into their countries' ways, leading them away from their Jewish roots, and ultimately threatening the continuity of Judaism in the Diaspora. A partial answer to these woes is to offer young people a free trip to Israel.

These philanthropists feel that an Israel experience will foster their Jewish identity and encourage them to support Jewish issues at home, marry other Jewish people, and raise their children as Jews. In fact, these leaders offer the students the opportunity to experience Israel for ten days, free, without obligation. They want it to be as easy as possible for these students to 'see Israel as the land of their ancestors and experience a country of other people just like them' (Taglit 1999: 1). Between December 1999-2001, more than 10,000 students from all over the world have participated in these trips.

What do the students think of the notion that they have a 'Birthright' to visit Israel? While the organizations involved do not offer an explanation of their program title, a 'Birthright' suggests much about the perceived importance and value of a trip to Israel for fostering Jewish identity and community outside Israel. The term implies that the participants have a historic responsibility or obligation to fulfill by visiting Israel. This perceived 'right' or duty, was an instrumental part of the argument for some early political and religious Zionists in establishing the physical state of Israel, and helped support subsequent Jewish national claims to the land (Berkowitz 1993, Cohen 1975, Hertzberg 1977, Vital 1975).

While claims to the land are contested at numerous historical moments on a variety of levels, the trip organizers use the term 'Birthright' to denote the Biblical roots of the claim to reinforce a Jewish connection to Israel. Through this claim, Birthright organizers encourage American Jewish youth to visit Israel for free, to strengthen the students' connection to Israel without the obligation of *Aliyah* (emigration to Israel).

Studying identity formation through travel is one way to look at how Jews define Israel as either a place, an identity, or both (Carter, Donald and Squires 1991: vii). Anthropological analyses regarding cultural nationalism and transnational identity studies are beginning to look at tourism and travel as grounds for academic inquiry, particularly in addressing intersections between Israel and diasporic Jewish communities (Bruner 1983, Dominguez 1989, Heilman 1994, Kirshenblatt-Gimblett 1998, Mittelberg 1999). Looking at students' photographs is a tangible means to look at these connections and perceptions from the students, and evaluate their connection to these ideas over time.

Previous Research

Assessments of students' Jewish identity and connection to Israel are made through changes in student responses over time through comparisons of pre-trip interviews, narrative accounts and interviews with their photographs after the trip. One innovative way to evaluate the ways students internalize their trip and express their sentiments about what they saw was to look closely at the photographs they took. Previous research by John and Malcolm Collier found that focused interviews on participants' photographs consistently reveal emotions and sentiments in terms of the questions presented, and remove emphasis and pressure from the informant (Collier and Collier 1986).

Previous research I conducted with first time American Jewish travelers to Israel revealed their stories subsequently led them to search for another image in their collections, and convey another anecdote, eventually making their way haphazardly through their photographs to convey their trip experiences. This communication cycle can reveal personal sentiments about travel and identity in a non-threatening manner. Participants' own sentiments about their own photographs lead them to reflect and project from their own experiences. In looking at travel photographs through a cycle of communication, scholars can gain insight in exploring the production, reception and consumption of visual media in many social and cultural contexts.

Georgetown University Students

This study focuses on nineteen Georgetown students that attended the June 2000 Birthright Israel trip: fourteen undergraduates, (including two graduating seniors), and five graduate students. Most undergraduate students are political science or international relations majors within the School of Foreign Service. The graduate students' backgrounds vary from law to public health to communications.

The Georgetown students generally consider themselves 'culturally Jewish', while most of their parents, if affiliated, are Conservative or Reform. Almost all the students' parents were born Jewish. They shared a variety of childhood Jewish experiences, both in and outside of synagogue affiliation, and use these memories to distinguish their 'cultural' Jewish identity from 'religious' practice. Each student reported receiving positive feedback from friends and family about going on a trip to Israel. Each mentioned they had always wanted to visit Israel and learn more about Judaism.

Ethnographic Strategy

I worked with Georgetown students' sentiments articulated over a nine-month period and the photographs they took during their trip as indicators of how students 'construct' their Jewish identity and connection to Israel. Students were asked to fill out pre-trip questionnaires, and were interviewed both individually and in small groups before the Birthright Israel trip. Students were observed and informally interviewed throughout their trip. Data was collected in written field notes, audio taped interviews and video taped educational programs. Each university represented on the trip provided staff for student groups. These university leaders, Hillel program managers and IEL tour company staff were also interviewed for program content before and throughout the ten-day trip.

When we returned to the US, I requested a set of photographs from each student. I also went to school events with students, conducted surveys and held interviews with them six months after our trip. During the final series of individual interviews I asked students to bring their photographs.

The Tour

The Birthright tour is constructed deliberately to connect the sites to both Biblical accounts and Jewish sites students may have previously heard or read about in other Jewish contexts. The students did not visit sites in the West Bank that also have Jewish significance, like Hebron or Qumran, where the Dead Sea Scrolls were found. In visiting the Old City of Jerusalem, groups walk through the Jewish Quarter via Jaffa Gate. They stop at the top of the Roman Cardo, pass the Huvrah synagogue and down to the Kotel pavilion to see the Western wall is a typical route for many Jewish and non-Jewish tours.

Each student had at least one photograph of the Western Wall and Dome of the Rock (Quabat eh-Sakhara). A total of 3% of all the photos used a similar angle of the Western Wall, taken from the steps above the security checkpoint when approaching the Kotel from the Jewish quarter. Another popular travel route the students participated in is to follow a particular series of events, like sleeping in the desert, waking up in the dark to climb Masada for the sunrise. Each student except one also had multiple photographs of the sunrise at Masada, approx. 4: 30 am (2% of photographs overall). He complained that so many people were taking pictures of the sunrise that he could not thoroughly enjoy it. Masada is Herod's famous hilltop desert fortress overlooking the Dead Sea. As described by Josepheus in The Jewish War, Masada was also the last stand for a small sect of Jews that committed suicide instead of becoming enslaved by the Romans.

These formulas allow tour guides in Israel to group sites within narratives that give symbolic power to the collective experience, connecting participants to events that others have participated in before them (Bourdieu 1990). They are repetitions of other family and friends tours, events outlined in pilgrim narratives from various Jewish texts, events and places discussed in Sunday school, Jewish camp, and travel guides about Israel. Anthropologists Daniel Boorstin, Nelson Graburn, Dean MacCannell and Edward Bruner have described these types of experiences as anticipated, fulfilled, and in some cases. On Birthright Israel, the students are participating in a history of Jewish travelers to Israel, and the program funding agencies hope that students will also be aware of this deliberate connection to the past throughout their visit.

Analysis

Over the nine months I spent with the Georgetown students, the ways they described what they saw influenced how they shared what they photographed. Each process as part of a circle of communication helped them define how they continue to struggle with how they see Israel and their Jewish identity.

Camaraderie was very important on this trip. Students could forget they were in Israel and had fun meeting other Jews their age. The bus they traveled on and the environment they created for themselves was a high priority. Hillel and Birthright's goals to create Jewish community were strongly supported by the actions and sentiments of the Georgetown students. The students' photographs alone reveal how important new and strong friendships on the trip were to them. Over two-thirds of each student's photographs were of other students and themselves (over 70% overall). It was incredibly important for the students to place themselves in the photos, either alone or with friends.

Over ten days, students averaged eleven or twelve pictures each day. Most students took a majority of their pictures in Jerusalem (22% of the photographs overall) over a four-day period (the middle 4 of 10 days). At the end of the trip I asked students to assess how many photographs they took on the trip, and where they took their most memorable photograph. On average, the students reported taking more pictures than they actually did. The students reported the most memorable photographs they took were of Masada, even though most of their pictures were actually of various sites in Jerusalem (22%), particularly in the Old City (9%).

Students took very few pictures of the sites they anticipated visiting. They took many more photographs of unexpected sites and activities. Like the Bedouin tent (4%) they slept in overnight before climbing Masada, riding camels (3.5%) and their Israeli tour guide. Previous research I conducted with first time visitors to Israel also had frequently photographed their Israeli tour guide.

Over half of the students said the 'desert' was their favorite place. They described the desert as a 'vastly different place' unlike any other they had seen before.

One student shared:

> The Negev desert [was my favorite]. It was so different from anything I had experienced, and so magical. I really realized there how lucky I was to be on such a trip and have that experience and I probably would never get to have an experience like this again.

Anticipated sites did not dictate how students took pictures. One student was most anticipating seeing the Golan Heights. He took the most pictures of the Golan (10) out of anyone in the group. However, he took more pictures of Masada (11) and Jerusalem (15). Another student was very interested to see Yad Vashem, but she only had one picture from the site. A second student, on the other hand, did not comment before the trip that she wanted to go to Yad Vashem, yet had taken nine pictures of different spots at the Holocaust memorial.

Of the sites we visited, each was not as easy to photograph, and could explain why the numbers of photographs at some sites was lower than others. Some students may have been told not to photograph religious sites or religious Jews out of respect for their prayer and privacy. That could explain why there are more photographs overall of camels than the Western wall, the most anticipated site among students. Also, ease of camera accessibility was a factor during specific activities. For example, it was much easier for students to take photographs of themselves in groups than while covering each other with mineral rich Dead Sea mud (less than 3%).

When Georgetown students were queried about the photographs they took and their favorite places together, I found that they did not necessarily take pictures of a place just because they found that site so enjoyable. Only in talking to each student about their experiences and comparing their individual comments with their photographs and conversations they had with each other after their trip were their thoughts about Jewish identity and community revealed.

Students described what they saw in terms of images they saw in their mind's eye, even if the photograph they were describing of a particular site we visited was not present, or did not even exist. One evening, students participated in an archaeological dig program. Jordan described to me a memory he had that 'Amy had found a cool artifact during our dig.' Upon searching through his album, he had six pictures of the archeological dig, but none of Amy or what she found. I knew what he meant, but he was disappointed he could not produce the image he was thinking of. Other students had pictures of the lamp she had found, including Amy!

Over time, individual students remembered more stories and more images and relayed more experiences. Their descriptions changed. The images they used to convey ideas were slightly different at different interviews. The ways students described their experiences to me were slightly different than when they described their trip to friends looking at their photographs. I was there, in Israel with them, and each student knew I had been to Israel before. Their tone and description was sometimes curt with me, and contained many 'Remember when ...' anecdotes. Friends looking at photographs and hearing of the trip for the first time, received much more thorough background information about the images. One student had a difficult time sharing his album with the 'mixed' crowd. He preferred to show friends the photographs and narrate while they browsed. He told Cory to look at the album on his own later that evening, and told me I could wait until the next week for his album, 'I have not seen everyone yet, and you were there, you remember.'

Students' photographs alone did not reflect trip expectations or emotional sentiments. Photographs, narratives and experiences together begin to reveal students' personal explorations of Jewish identity, participation in a Jewish community in the United States, and relationship with Israel. Students did not photograph the sites they anticipated the most, or their favorite sites. Generally, students photographed themselves and their friends at as many places as possible, both as proof of their experience and to emphasize the importance of social networks.

Individual experiences, cultural baggage, historical circumstances, and an unencumbered sense of 'camera culture' all play a role in tourist photography (MacCannell 1989). First time travelers to Israel all have a slightly different threshold, or spectrum of what we as individuals and as members of a larger Jewish community consider to be 'authentic.' Eric Cohen found that 'in the Holy Land, early photographers' conceptions of 'authenticity' were biased by the country's biblical past and their own cultural predispositions' (Cohen 1992: 227).

Students' photographs cannot be separated from sentiments about their trip experiences and the incorporation of these experiences into their future participation in a Jewish community. Their thoughts are reinforced, remembered and changed through the images. The images influence how they remember their trip and communicate about it to others.

The goal of this research is to examine this experience for nineteen specific students. These participants gave many responses laden with supporters' expectations. However, they did not always feel the trip was fun, amazing, or impressive. At many points the Georgetown students felt they were seeing a 'contrived' Israel, not the 'whole Israel' and only a 'Jewish' Israel. The Georgetown students were consciously aware they were not seeing the whole Israel, or possible alternatives, including Palestinian values or some Israeli interpretations on the politics of peace in 2000.

I was surprised to learn the Birthright trip devoted a very small amount of time to any Israeli–Arab voices and no time at all to Palestinian voices. Birthright operates within the symbols, myths and debates inherent in the literature of American Jewish education and tourism. The structure of the trip maintained focus on memories and experiences of Zionism. During the trip, students expressed interest in wanting to learn more about conflicts over land ownership in this region. These are impossible topics to overlook in correlating Jewish identity with a connection to the physical state of Israel. These students were over zealously presented with a more Zionist side of the politics of Israel's borders. The places they visited also remained inside those 'borders.'

Two women have developed a new program that addresses these complex issues. Initiated in 2003, Birthright Unplugged hosts Jewish participants, including college students, on six day trips through parts of Israel and the West Bank. This program asserts that it addresses issues related to Palestinian voices, political climate and concerns of the region. Comparative longitudinal research comparing Jewish identity, cultural ties with Israel, photographs and narratives from these two programs is ground for future inquiry.

Constructing Jewish identity is more than memory and sites. Students' photographs combined with their sentiments over time showed growing connections between what they thought about Israel and being Jewish before and after the trip. After the trip, Georgetown students combine being Jewish, and thinking about 'Israel.' In talking about their experiences, showing others their photographs, and reflecting on their trip six months later, students express much more conviction about what Israel is about, and significantly less conviction about what being Jewish is about. However, where once being Jewish and Israel may have been separate concepts, students' comments six months later reveal these concepts are now intertwined, both part of the larger ideas of Jewish identity and community. Neither of these sentiments are mutually exclusive for these students six months after their trip, but inform and are informed by each other.

What difference does it make to look at students' photographs? From looking at students' photographs, it could be interpreted that students brought with them in their cultural baggage the insight to compliment the moments they experienced by photographing poses, angles, and sites. It is possible that student 'feelings' could be recreated in the photos. Interviews with the Georgetown students describing their own photographs and their own experiences indicate they often described how they felt or what was happening before, during and after the image was taken. Looking at photographs with the students led them to talk about events that lacked any photography. The photographs and stories created a context that changed over time, and can be indicators of future changes in behavior and attitude toward Israel and their Jewish identity.

References

—— (1990), The Council of Jewish Federations. National Jewish Population Survey. <http: //web.gc.cuny.edu/dept/cjstu/demhig.htm>. Accessed 15 March 2001.

—— (1999), Taglit/ Birthright International (ed.) Program Guidelines for Taglit/ Birthright Israel International.

—— (2000), *Birthright Israel Background Summary* (June 2000).

—— (2006), *Birthright Unplugged*. Itinerary and Program Goals. <http: //www. brithrightunplugged.org>. Accessed 10 May 2003.

Berkowitz, M. (2000), *The Jewish Self-image in the West*, NYU Press, New York, NY.

Berkowitz, M. (1993), *Zionist Culture and West European Jewry Before the First World War*, The University of North Carolina Press, Chapel Hill, NC.

Boorstin, D.J. (1971), *The Image: A Guide to Pseudo-Events in America*, Athenaeum Press, New York.

Bourdieu, P. (1990), *Language and Symbolic Power*, Polity Press, Cambridge.

Bowman, G. (1996), 'Passion, Power and Politics in a Palestinian Tourist Market', in T. Selwyn (ed.), *The Tourist Image: Myths and Myth Making in Tourism*, John Wiley, New York, pp. 83-104.

Bowman, G. (1994), 'Tales of Lost Land: Palestinian Identity and the Formation of National Consciousness', in E. Carter, J. Donald and J. Squires (eds), *Space and Place: Theories of Identity and Location*, Lawrence and Wishart, London, pp. 73-100.

Bowman, G. (1992), 'Pilgrim Narratives of Jerusalem and the Holy Land: A Study in Ideological Distortion', in A. Morinis (ed.), *Sacred Journeys: the Anthropology of Pilgrimage*, Greenwood Press, Westport, CT, pp. 149-168.

Bruner, E. M. (1983), 'Dialogic Narration and the Paradoxes of Masada', in *Text, Play and Story: The Construction and Reconstruction of Self and Society*, Proceedings of the American Ethnological Society, pp. 56-79.

Bruner, E. M. (1995), 'The Ethnographer/Tourist in Indonesia', in M.-F. Lanfant, J.B. Allcock, and E.M. Bruner (eds), *International Tourism: Identity and Change*, Sage Publications, London, pp. 143-158.

Carter, E., Donald J. and Squires J. (1991), 'Introduction', in *Space and Place: Theories of Identity and Location*, Lawrence and Wishart, London, pp. vii-2.

Chalfen, R. (1987), 'Tourist Photography: Camera Recreation', in *Snapshot Versions of Life*, Popular Press, Bowling Green University, OH, pp. 100-118.

Chalfen, R. (1980), 'Tourist Photography', *Afterimage*, Summer, pp. 26-29.

Cobley, P. and Litza J. (1995), *Introducing Semiotics*, Totem Books, New York.

Cohen, E. (1992), 'Pilgrimage and Tourism: Convergence and Divergence', in A. Morinis (ed.), *Sacred Journeys: Anthropology of Pilgrimage*, Greenwood Press, Westport, pp. 47-64.

Cohen, E., Yeshayahu N. and Uri A. (1992), 'Stranger-Local Interaction in Photography', *Annals of Tourism Research*, Vol. 19, No. 2, pp. 213-233.

Collier, J. Jr. and Collier, M. (1986), *Visual Anthropology*, University of New Mexico Press, Santa Fe, NM.

Dominguez, V. (1989), *People as Subject, People as Object: Selfhood and Peoplehood in Contemporary Israel*, University of Wisconsin Press, Madison, WI.

Dominguez, V. (1995), 'Invoking Racism in the Public Sphere: Two Takes on National Self-Criticism', *Identities*, Vol. 1, No. 4, pp. 325-346.

Graburn, N.H.H. (1989), 'Tourism: The Sacred Journey', in V. Smith (ed), *Hosts and Guests: The Anthropology of Tourism,* 2nd Ed, University of Pennsylvania, Philadelphia, PA, pp. 21-36.

Heilman, S. (1994), *A Walker in Jerusalem*, Jewish Publication Society, Philadelphia.

Hertzberg, A. (ed) (1977), *The Zionist Idea: A Historical Analysis and Reader*, Atheneum Press, New York, NY.

Kirshenblatt-Gimblett, B. (1998), *Destination Culture: Tourism, Museums, and Heritage*, University of California Press, Berkeley, CA.

MacCannell, D. (1989), *The Tourist: A New Theory of the Leisure Class*, Shocken Books, New York.

Michaels, E. (1991), 'A Model of Teleported Texts (with Reference to Aboriginal Television)', *Visual Anthropology*, No. 4, pp. 301-323.

Mittelberg, D. (1998), *The Israel Connection and American Jews*, Praeger Press, Westport, CT.

Perin, C. (1992), 'The Communicative Circle: Museums and Communities', in I. Karp, C. Mullen Kreamer, and S. D. Levine (eds), *Museums and Communities: The Politics of Public Culture*, Smithsonian Press, Washington, DC, pp. 182-220.

Sobel, R. (2002), 'Imagineering Israel: (Re)Constructing History Through Travel Photographs', in L.J. Greenspoon and R.A. Simkins (eds), *A Land Flowing with Milk and Honey: Visions of Israel from Biblical to Modern Times*, Nebraska University Press, Omaha, NB, pp. 179-199.

Vital, D. (1975), *The Origins of Zionism*, Clarendon Press, Oxford.

Chapter 14

The Purloined Eye: Revisiting the Tourist Gaze from a Phenomenological Perspective[1]

Marie-Françoise Lanfant

Nous avons braqué sur la durée un oeil
qui l'a rendue durante.[2]

(Claudel, 1946: 191)

Introduction

Some time ago I was invited to present a paper at an international conference in Sheffield in the United Kingdom. The title of this conference: *Tourism and Photography: Still Visions, Changing Lives* was to re-open for me a question which John Urry had announced about ten years earlier in his book *The Tourist Gaze* (1990). The expression 'tourist gaze' had immediately caught my attention when a copy of the book was given to me during a conference I co-convened in Nice, France in 1992 (Lanfant 1992). The terminology used was by then not new. The expression of the '*regard touristique*' (imperfectly translated as the 'tourist gaze') had been circulating for a while in discussions. The research team at the French National Centre for Scientific Research (CNRS) which I had directed between 1975 and 1997, had consistently discussed the issues which John Urry raised in his book notably; the theses by Boorstin on 'pseudo-events' (Boorstin 1961); de Debord (1983) on the 'society of spectacle'; Baudrillard (1972, 1981) on the political economy of the sign and; MacCannell (1976) on the structural semiotics of tourism.

Questions related to the gaze by Western tourists upon so-called developing countries were at the heart of many debates and concerns that animated our international research networks, as well as the work of our colleagues carrying out ethnographic fieldwork in different developing countries. These different works were threaded through a series of international meetings during the 1980s and 1990s, and the World Congresses of the International Sociological Association, particularly in 1990 in Madrid and in 1994 in Bielefeld where John Urry presented a paper on visual consumption (Lanfant 1992, 1995). In 1998, as the Vice-president and organizer of a meeting within the World Congress of Sociology in Montreal, I organized a round table discussion with John Urry and Dean MacCannell, in order

---

1 Translated from French by David Picard.
2 'We have cast an eye on the passing of time that made it enduring'.

for both to explain and confront their respective approaches. This, in my view, proved to be rather unsatisfactory in that it failed to interrogate the essence of the gaze and situate it in some of the philosophical and psycho-analytical concepts that had preceded it. This chapter attempts to re-locate the idea of the 'gaze' in a wider body of work from the French intellectual milieus of the Post-War period which were heavily influenced by Husserl's notion of phenomenology. I explore the gaze as a departure point for a reflection on the complexity of what is shown, what is 'given to see' and what is actually seen in the field of tourism; an approach that goes beyond the principles of scopic vision and, the epistemology of the science of the gaze.

Questioning the Gaze

If I understand the meaning of the term in English, to gaze is not just seeing, but seeing something or somebody intently. It is the gaze of the observer, the gaze of someone looking at something or at someone as if 'from above'. To gaze comprises both a visual and a scopic dimension of *le regard*. But what exactly is meant by the expression 'tourist gaze'? Is it a certain way of seeing, or specifically a touristic regard? Is it the gaze of the tourist or of the tourists? Is it the perception of something made visible to tourists only or made visible to everyone? Is it a mirroring game between the gazer and the gazed upon? Is it the perceptive field of a specific regard? Is it a quality ascribed to the things seen? Is it a gaze above the 'world that one can see'? Is it a gaze by a professor on a particular subject?

In John Urry's discourse, it appears as an amalgam of all of these interpretations. Urry goes further and claims, though a little vaguely, that the tourist gaze is a 'social construction' (1990: 1-2). He writes about visual *consumption* hence suggesting a gaze that 'eats with the eyes'. Yet for Urry the tourist gaze is not a simple metaphor but takes on the value of a concept. It is a nexus of significations situated at the crossroads of different analytical plans of tourism as a social and cultural phenomenon which is leading to a major question: is the tourist gaze representative of a new mutation of the way of representing the world? Or, to ask it more clearly: is the tourist gaze a representative of contemporary social representation?

The Tourist Gaze (1990) appears to work with the concept without clear definition. This may have several reasons, the most important one being that one cannot approach the question of the gaze directly. This can be done by trying to make the perceived (*percipi*) world significant, by interrogating the percipients (the seeing subject) or by analysing the semiotic structure of social or physical environments. Tourism has familiarized us with expressions including 'giving to see' (*donner à voir*), 'worth seeing' or 'must sees'. The impression from the original *Tourist Gaze* is that touristic representations of the world are forged through the visit of places made visible to them; places shown and conditioned by the means of tourism promotion and in many ways artificially created to flatter the curiosity of tourists and incite their desire to see. As a result, the world tourists see becomes an embellished, illuminated and unreal artifice.

However, let us admit that whoever we may be, tourist or anthropologist, we share a common belief in the reality of what we see. Things that we see appear well real to us. Following Merleau-Ponty (1964), we can say that 'this belief is founded on a deep sitting base of silent opinions implicated in our life. Yet this belief is strange in that, for the philosopher, when he/she tries to articulate it into a thesis or conceptual outline, when he/she asks what that is to see, to see things of the world, he/she enters into a labyrinth of difficulties and contradictions' (Merleau-Ponty, 1964: 17, *).[3]

In his introduction to his 1990 text of *The Tourist Gaze*, John Urry suggests an indication of the theoretical frame in which the concept of the tourist gaze would be operational. He makes reference to Michel Foucault suggesting that he intends to do with the tourist gaze what Foucault had done with the medical gaze in his *Naissance de la clinique* (Foucault 1963). Yet Urry's quotation of Foucault's book significantly remains incomplete. The exact title of Foucault's work is *Naissance de la clinique, Archéologie du Regard Médical*. However, *The Tourist Gaze* does not significantly engage with Foucault's problematic, approach and method. The analogy between the tourist gaze and the medical gaze is never fully explained (though Urry acknowledges this in the additional chapter of his 2002 edition of the book and makes reference to this critique levied by Keith Hollinshead, 2000). Certainly, Urry talks about visual perception and visual consumption and brings together established notions such as perception and consumption. However, instead of providing illuminating clues, these allusive references reinforce our perplexity precisely because they relate to theoretical backdrops that the content of this book does not consider in an adequate way.

In this text, and in the context of approaching the range of fundamental relationships between tourism and photography, I would like to suggest a set of elements that specify the type of approach the use of the Foucauldian notion of *gaze* brings with it. I believe that we need to go back to the premises and ask whether in our field of research this concept has a heuristic value, with or without Foucault. The specific theme of *Tourism and Photography* discussed in this text offers here a good occasion to reconsider the latent questions which still remain many years after the first edition of John Urry's book. Linking tourism and photography certainly offers a good perspective to explore the scopic structure of the contemporary subject of whom the tourist may be considered as a pure representative. In this sense I situate my approach within the overarching questionings of this present volume which stem from the clear observation that the taking of photographs is a central part of tourist behaviour. Thus, we can ask: why do tourists take photographs? Why do they frame certain things in the world whereas others seem to remain 'anaesthetic'? Do tourists 'shoot' photographs or do they 'capture' the 'magical' power of the instant? What are the implications for the analysis of tourism as a

---

3   Asterisk (*) denotes quotation translated into English by David Picard

'sacred time-space' embedded with and opposed to everyday life? The approach of photography as part of the tourist experience casts a new light on concepts regularly used by researchers interested in the tourism field. At the same time, it introduces new conceptual and methodological frameworks borrowed from visual anthropology, sociology, psychoanalysis, cognitive psychology, political science and philosophy' (Robinson and Picard 2003).

Background of the *Tourist Gaze*: Returning to Sources

One could possibly better understand the actual objective of John Urry's book by reading the first edition of Michel Foucault's *Naissance de la clinique: une archéologie du regard médical* (Foucault 1963). Foucault chose the term '*gaze*' to describe the extension of a discursive corpus. Effectively, since the post-war period of the 1950's, the *gaze* has become one of the most obstinately studied concepts within the Paris based French academic world. It was addressed by great thinkers such as Jean-Paul Sartre, George Bataille, Maurice Merleau-Ponty and Jacques Lacan. All these authors entertained close working relationships. The actual places they were teaching in were close to each other and institutionally coordinated. There was a constant flow of meetings and exchanges, a close interaction between the outputs of each of them. Michel Foucault's approach emerges within this particular context. Then a student at the *École Normale Supérieure* (ENS), as were before him Jean-Paul Sartre and Merleau-Ponty, he benefited from the ferment marking this particular intellectual environment. I believe that his work is meaningful only when resituated within these specific affiliations. Seen that the author of the *Tourist Gaze* takes Michel Foucault as the guide for his approach, it is not uninteresting to reassemble the threads Foucault used to construct himself the concept of the gaze.

It is not, however, my intention within the limited space of this chapter to present in detail the theses of the authors I will quote. I will restrict myself here to indicating the important steps of a reflection on the concept of the gaze that have become sign-posts for the later development in our own research. In my view, the authors I will quote form a reference framework fundamental to the construction of an analytical approach of the phenomenon we are interested in; *The Tourist Gaze*.

Foucault was trained at the ENS in Paris and his academic work would remain deeply influenced by the teachings dispensed in this *Grande École*, specifically Gaston Bachelard's work on epistemological rupture and the new scientific spirit and Georges Canguilem's research on ideology and rationality in the life sciences. He was particularly interested in the question of the gaze; a question other French authors, many of whom have acquired reputations that go far beyond the scope of the Parisian intellectual circles of the 1950s and 1960s, have progressively continued to develop. In many ways, this question has become the object of an essential interrogation at the intersection of psychoanalysis, phenomenology and the sociology of knowledge.

Husserl and the Origins of the Phenomenology of the Gaze

The initial inspiration here comes from Edmund Husserl. Husserl's conception of phenomenology initiated a thoroughly new way of dealing with the philosophical and epistemological problems of knowledge, the cogito (the subject of *connaissance*) and scientific objectivity. I will not resume Husserl's message here in a couple of lines. For a good initiation to phenomenology, I would suggest the reader to study the translation of Husserl's original texts, such as *The Idea of Phenomenology: Five Lessons Pronounced at the University of Heidelberg in 1907* (Husserl 1985), and those by one of his students, Martin Heidegger. Heidegger assumed the continuity of Husserl's teaching at the University of Freiburg after Husserl, being prosecuted by the Nazi regime for his Jewish ancestry, had to leave his academic chair. Heidegger's major publication, *Sein und Zeit*, first published in 1927, includes a couple of enlightening pages on the idea of phenomenology (Heiddegger 1986: 53-66). What we should in particular remember is what Husserl himself said about phenomenology 'designating above all a method and an attitude about thinking'. (Husserl 1985: 41).

With regard to this method, one primary rule should be retained: to carry out a rigorous description of the lived experience of the subject in the world of things and others before any form of speculation on human nature and the 'laws' governing the social and physical environments in which it is evolving. This research perspective has generated a radical change of the way of recollecting, describing and conceiving the facts that characterise the specificity of disciplines in the humanities and social sciences. Terms such as 'lived experience' (*Erlebnis*), 'subject', 'ego', 'alter-ego', 'otherness' and 'alterity', 'intentionality of consciousness', 'subjectivation', 'intersubjectivity', 'subject in situation', 'one's own body' (*le corps propre*), 'in between brackets', 'objectivation', the 'site' and the 'world' (*le site et le monde*), 'reality'/'quasi reality', and so on, have entered the mainstream discourse of our disciplines, namely in psychology, sociology and anthropology. To return back to the specific theme of the gaze we are interested in here, it suits to recall that Husserl accorded perception and sight an original role in the constitution of knowledge. He hence took on one of the fundamental problems humanity has been tantalized by ever since it started practicing philosophy: the question about human nature and the relativity of knowledge. This question had earlier been raised by the English thinkers Locke, Hume, Berkeley and their heirs, but had remained in suspense. By developing a radically new method, Husserl allowed philosophers and social scientists to re-approach this question from a new perspective.

In the Footsteps of Phenomenological Epistemology

An analysis of the gaze directly based on the lecture of Husserl's writings would take us too far within the limited space of this article. However, it seems important to me to link the concept of 'tourist gaze' to the sum of works carried out in France and elsewhere based on the method advocated by Husserl. In fact, the most original essays about the problem we are interest in have been written by authors whose own reflections were influenced by Husserl's writing.

An important essay was written by Eugen Fink, one of Husserl's disciples. In 1927, the philosophical faculty of the University of Freiburg suggested a dissertation prize competition on a theme very close to the subject of this chapter. It asked authors 'to address, distinguish between and examine through a pure phenomenological analysis the psychic phenomena defined by the ambiguous (*vieldeutig*) expressions of "to think as if" (*sich denken als ob*), "just to imagine something" (*sich etwas bloss vorstellen*), and to "fantasise" (*phantasieren*)'. Eugen Fink won this competition and developed his initial text, *Representation and Image*, into his dissertation, submitted two years later, under the direction of Husserl. Fink's thesis presents a crucial interest for us in the way it distinguishes the notion of 'image' from the notion of 'representation' (Fink 1966). In phenomenology, these are two modes for a thing to make itself manifest, to make itself visible. Representation as analysed in philosophy, is what comes at the place of the thing, in form of ideas, beliefs, fictions, etc. The analysis of representations leads us to the moving universe of the symbolic function structured by language. The image is of a different essence; it is the reproduction of the thing in a mirror, in the eye of a viewer, on a screen, on a photographic film or digital chip. The image witnesses an immediate contact between the body and the world of things.

Research taking up Husserl's new approach only developed slowly in France, mainly due to the rarity of publications and the obstacles of the language. In 1929, the Institute for German Studies and the French Society of Philosophy jointly invited Edmond Husserl to animate a series of conferences at the Sorbonne in Paris. Initially presented in German, these conferences were later translated into French and published in 1947, with the title *Méditations Cartésiennes* (translated into English as *Cartesian Meditations*, 1999). During a study trip through Germany, at the eve of the Second World War, Jean-Paul Sartre developed a passion for the philosophy of Husserl and acquainted with the philosophy of Heidegger (*Zeit und Sein*, 1927, translated into English as *Time and Being*, 1962).

Jean-Paul Sartre's masterwork: *L'être et le néant*, published in France in 1943 during the German occupation (translated into English as *Being and Nothingness* [1957] 2007), clearly transpires this influence. It includes eighty pages on the theme of the *gaze*, greeted some years later by Jacques Lacan, the psychoanalyst, in his 1964-65 seminar as being 'the most brilliant that the existentialist philosopher had written' (Lacan 1973: 70). For Sartre, 'the gaze implies the existence of the other (in the self), before any verbal exchange between the other and the self' (Sartre

1943: 319; *). This, it seems, corresponds to the case of the tourist gaze. What does it mean for the self to 'be seen'? As Sartre continues:

> The gaze of the other reveals me in my pride of being signified as being seen, or in my shame of being looked at. I am for the other an object of observation. My situation, my environment become those of the other. Consequence: seeing objects empowered to see me induces the sliding of the entire solid (*figé*) universe that is mine. With the gaze of the other, I no longer master the situation; or it rather escapes me. (319; *).

At the same period, during the years of the Second World War, Merleau-Ponty devoted himself to a rigorous reading of Edmund Husserl. In 1945, Gallimard, Paris published Merleau-Ponty's *Phénoménologie de la perception* (transl. into English as *Phenomenology of Perception*, 1962). Sartre and Merleau-Ponty were friends working together as editors of the *Revue des Temps Modernes*. In this context, their intellectual exchanges were quotidian and would last until their fracture, provoked by different attitudes towards Marxist regimes. From 1949 to 1953, Merleau-Ponty held the chair of General Psychology at the Sorbonne. His research was directly informed by the reading of Husserl's and his students' works. In his *Phénoménologie de la Perception*, Merleau-Ponty explicates the perception of the world of things, of others and of one's own body and based on the lived experience of subjects in situation dissects the question of what is it 'to see'? As Merleau-Ponty (1945) explains: 'Through phenomenological reflection, I find VISION not as a "thought to see" (*pensée de voir*), according to Descartes, but as a GAZE (*REGARD*) having a hold over the visible world.' (404; *)

Phenomenology thus takes us back to the starting point of the actual formation of vision, to the centrality of the eye as a physiological organ, emerged from the *Leib*, from the flesh of the world, the unnamed matter of life. According to an explication by Husserl (1982):

> the analysis of sensitivity predominates as soon as the "Leib" is presented as the formatting instance of constitution [...]. The original experience is one of interlacing, of the intrication of the flesh and the world: that is the sense of "Umwelt", identity of site and world [...]. The flesh – LEIB – is what exceeds the body as "Körper" [...]. The theory of the body as Leib breaks with the metaphysical oppositions of the object and the subject, of the inside and the outside, the self and the other. (14; *).

Merleau-Ponty (1945) repeats this idea in his own words:

> Because we have a surface in contact with the world, a perpetual deep-rootedness in the world, all knowledge installs itself in the horizons opened by perception. [...] We need to conceive the perspectives and the point of view as our insertion into the world and perception no longer as a constitution of a true object, but as

our inherence to things. […] Perception only is perception of something when at the same time being imperception of a horizon or a background it implies, but that it does not thematise. (402; *).

From this point Merleau-Ponty continues:

It can't be possible any longer to isolate, in my physiological representation of the phenomenon the retinal images and their cerebral correspondent, actual and virtual in which they appear. One will no longer be able to realise under the name of psychic images the discontinued views that correspond to consecutive retinal images; neither introduce "an inspection of the spirit" that resituates the object through deforming perspectives. (402; *)

Commenting on Merleau-Ponty's book, Lacan, in his *Les quatre concepts de la psychanalyse* (1973) situated its principal perspective and recognized that:

Phenomenologists have managed to assert with most clarity that perception is not located in the self but that it is "upon" the objects it apprehends, in contrary to the conception of the naive conscience assuming that representation belongs to the self of the observer. We are facing a philosopher who seizes something that is one of the essential correlates of consciousness and its relation to the unconscious. (68;*)

In this respect, one will notice the convergences operated from then on between the *in situ* observations of young children and animals made by the new psychologists and the outcomes generated by clinical psychoanalysis – witnessed in particular by the work of Sigmund Freud re-read and explicated by Jacques Lacan in his seminaries run uninterruptedly between 1953-54 and 1978-79.

Merleau-Ponty's publications (1968, 2003) teem with multiple experiences he gathered about the behaviour of children and adults in specific or experimental conditions and which served him to elaborate his conception of perception. For example, he exposed and critiqued the description that the most prominent psychologists would make of the speculare image, that is to say the image the baby (or young animal) receives of itself by looking in a mirror. This situation – the recognition by the subject of his/her own body reflected in the mirror – was an experimental protocol used by different psychological schools of thinking to analyse processes of a subject becoming conscious of the self and the world. At the same time, psychoanalysis elaborated a conception of the relation between the desire of the 'Other' and narcissism, starting from the moment the young child discovers its image in a mirror while its mother is watching. These observations lead Jacques Lacan (1966, 1982) to propose his theory of the mirror stage which would remain the starting point for all the ulterior development of his analysis of the gaze as 'object A'.[4]

In 1961, Merleau-Ponty, then Professor at the *College de France*, died suddenly. His posthumous publication, *Le visible et l'invisible*, put together by Claude Lefort in 1964 (translated into English as *The Visible and the Invisible*, 1969), provoked a thriving return of the gaze as an object of sociological study. From this point, Durkheim's initial approach of apprehending social facts as things was not, or no longer, evident. This early rule of the sociological method was critiqued and progressively replaced by a new methodological and epistemological position establishing the observing subject within the observation framework (for example by participant observation methods used in ethnographic fieldwork). Through the engagement of his or her own body and subjectivity, the observer of social facts became thoroughly connected to the object of observation and made visible his or her apprehension of these facts. While the subject of science lay in the object of sociology, one had to ask whether sociology was therefore 'scientific' in a strict sense, or whether it is more or less 'scientific', a 'soft' science.

Jacques Lacan, one of Merlau-Ponty's friends, immediately commented on the latter's posthumous text. In his 1964-65 seminar, *Les quatre concepts de la psychanalyse* (1973 and translated into English as *The Four Fundamental Concepts of Psycho-Analysis* in 1998), Lacan repeatedly addressed the question of the gaze by asking about 'how we can situate the conscience from the perspective of the unconscious' (51). By then, Lacan, as explained earlier, had long been situating the gaze as a primordial object of psychoanalysis as 'object A'. Lacan explains that 'the introduction of the object a makes obsolete any type of conception that reduces the gaze [*regard*] to a scopic function within a structure of vision' (Lacan 1973: 63-109). As with Freud, what retains Lacan's attention is the dialectic of desire working in the function of seeing and observing. The discovery of the unconscious subverts any conception based on the idea of a correspondence between what is shown, what is given to see (*donné à voir*) and what is seen. In Lacan's words: "In general terms, the relation between the gaze [*regard*] and what one wants to see is deluded. The subject presents himself differently from what he is, and what he is given to see is not what he desires to see' (96).

---

4   In the Lacanian algebra, 'object A' designates the algorithm of desire. To better understand this notion, it seems helpful to go back to Lacan's seminars. Since his first conjectures based on empirical studies, in particular his 1936 theory of the mirror stage, he would persistently come back to his seminars of 1954-55: *Le Moi dans la théorie de Freud et dans la théorie de la Psychanalyse* (Lacan 1978); of 1964-65: *Les quatre concepts de la psychanalyse* (Lacan 1973); of 1965-66: *L'Objet de la psychanalyse* (Lacan 2005).

The Project of Michel Foucault

It seems to me that Michel Foucault's project, despite the inspirations it took from the vocabulary of these famed thinkers exposed above, is situated in a different order of preoccupations. What was Michel Foucault's aim when writing the *Birth of the Clinic*? According to Foucault, this book (with its subtitle *Archaeology of the Medical Gaze*) belongs to the history of ideas. For its author, it aimed at explaining 'the historic moment in which medicine becomes a positive science discharged from metaphysical presuppositions related to the human body, disease, death, and other areas subjugated, by language, to specific ways of seeing and saying' (Foucault 2000: v). For Foucault, this historic moment in the history of medicine is an indicator for a wider epistemological rupture attaining the fields of knowledge and cognisance. In this sense, his study is building on the works developed by Gaston Bachelard (1949, 1985) and George Canguilhem (1977, 1991).

What is important here is to understand the hypotheses underlying this approach. For Foucault it was not only about studying ideas and practices from a historical perspective, but the interrogation, at the same time, of the mental structures and psychic processes that support belief systems, representations, values and associated practices in an approach he called 'archaeology of knowledge', an expression appropriated from Merleau-Ponty, who himself received it through his reading of Husserl (Merleau-Ponty 1945: 11-13). However, revising the relationship between the enunciable and the visible, Foucault's approach progressively differentiates itself from the principles of phenomenology (Deleuze 1986). From this point, *The Birth of the Clinic* (Foucault 1963) needs to be seen as part of a wider project. It comes as a complement to Foucault's previous book *History of Madness* (2005) and it preceded his *Archaeology of Knowledge* (Foucault 2002) and *The Order of Things: Archaeology of the Human Sciences* (Foucault 2001). During his inaugural lecture at the *Collège de France* in 1966, Foucault retrospectively resituated these diverse publications within the framework of his original project. Presenting his future research programme, he explained that:

> I am placing myself at the turn of the XVIth and XVIIth centuries when especially in England a science of the gaze [regard], of the observation, of the statement appears, a certain natural philosophy without doubt inseparable from the implementation of new political structures, inseparable also from religious ideology: a new form, certainly, of the will for knowledge. (Foucault 1971: 64; *)

Foucault's statement invokes George Berkeley's early 18th century philosophy (qualified as 'immaterialism' by its author) supposing a natural equation between words and things: In Berkeley's words: 'It is customary to call written words and the things they signify by the same name: for words not being regarded in their own nature, or otherwise than as they are marks of things, it had been superfluous, and beside the design of language, to have given them names distinct from those of the things marked by them' (Berkeley 1732: 140). Where does Berkeley's theory of vision take us? The world that I am seeing, in which I am implicated, is sustained and authenticated by an 'other'. In Berkeley's philosophy, this 'other' is God. For him, the perception of the world hence leads to God. The signification of the world of things is suggested by God, omnivoyant, omnipresent, and omniscient in all matters. By God alone.

We see how much this theory of vision presents analogies with the concept of the 'tourist gaze' as conceived by Urry. The ghost of Berkeley remains in our spirits. To which degree does the 'tourist gaze' pursue this laudation of a non-identified Grand Other? If it is not God, who then is the 'other' who incites rejoice from the spectacle of an embellished world, intended at the enjoyment and delight of tourists?

Returning to the Birth of the Clinic

In the first lines of the preface of the 6th edition of the *Naissance de la clinique* (translated into English as *Birth of the Clinic*, 2000), Foucault announces that 'it is a matter in this book of space, language and death, it is a matter of gaze' (1963: V; *). Whose gaze is it? This introduction seems odd in that it relates three words whose connotations do not necessarily articulate what they denote joined together in an elliptical way by the notion of the gaze. It is eventually comprehensible when referring to the exercise of medicine in that the change occurs in the manner of treating illness from the moment the doctor leans over the sick, observing *in visu*, with a meticulous gaze, noting and commenting on what he sees. Foucault explains that 'at the beginning of the 19th century, doctors started describing what for centuries had remained below the threshold of the visible and the enunciable; what was fundamentally invisible suddenly offers itself to the clarity of the gaze' (199; *). He continues:

> One has the impression that for the first time in millenniums, doctors, freed of theories and chimeras have consented to approach themselves, with a non-informed eye the object of their experience, [...] to say what they see and also to give to see [donner à voir] by saying what they see. [...] The relation between the visible and the invisible has changed its structure and makes visible through the gaze [regard] and language what was beyond their domain. A new alliance was made between words and things, making visible [faisant voir] and making say – sometimes in such a truly naïve discourse that it appears like situating itself at a more archaic level of rationality. (200; *).

This more archaic level of rationality is what Foucault, or rather his publisher, called the 'archaeology of knowledge'. The term 'arche' is itself engulfed with incertitude, able to lead us to the obscure and repressed regions of unconscious knowledge. Yet, for the time being, let us consider medicine, having left its chimeras and become a scientific approach to illness that takes its source in the technique of observation and discourse, of a language applied to this practice. The medical gaze is a specific gaze. It is a gaze that observes, a gaze that listens;

> ...that restrains itself from intervention: it is mute and without gesture in what it offers. The knowledge of a thing is deduced from the structure of a perceptive experience between observers and observed, in which the observed are put into an object position. The correlate of the observation is never the invisible, but always the immediately visible. There is nothing for it to hide in what it is giving (Foucault, 1963: 109; *).

This presupposition leads to a (scientific) esotericism, the belief that the manifestation of a thing leads to its truth, the gaze as a way of initiation to the truth of things. The experience between words and spectacle, and vice and versa, is based upon a formidable postulate; 'Above all efforts of clinical knowledge floats the big myth of the pure Gaze, pure Language, the eye that talks by itself'. (Foucault 1963: 109; *). Foucault also calls the sovereign power of the gaze, the sovereignty of the gaze and passionately concludes that this 'is the advent of scientific positivism' (200; *).

Can we talk in these terms about the tourist gaze? I think this question merits further development. When reading *Birth of the Clinic*, one gets the impression that the question of the gaze has been retracted. Michel Foucault recognised this himself. In the preface of the 6th edition, he self-critically reconsiders that when he was initially writing his book, he had invoked 'a medical gaze that supposed the unitary form of a supposedly fixed subject in relation to an objective field' (Deleuze 1986: 22). In this 6th edition, the subtitle 'Archaeology of the Clinical Gaze' had disappeared, for a reason that remains undisclosed by the author. Michel Foucault seemed more fond about the primacy of the statement concerning ways of seeing and perceiving.

In order to situate the tourist gaze within the line of research developed by Michel Foucault, it seems helpful to refer to another of his books, *Les Mots et les Choses*, with its subtitle *Archaeologie des sciences humaines*, in which he interrogates the coherence between the theory of representations and the one of language that had existed throughout the classical age (Foucault, [1966] 1998). It seems necessary to focus in detail on the chapters that interest specifically our subject of the tourist gaze, in particular the first in which he tests his own science of the gaze through an analysis of the painting *Les suivantes* by Velasquez,[5] to illustrate the thesis he asserts in *Les mots et les choses* (Foucault [1966] 1998: 19-31). In this text, Foucault comments forcefully, point after point, the different elements of the painting, focusing especially on the *jeux de regards* (literally 'the play of eyes casted over') between the figures painted on the canvas. In the conclusion of his chapter, he explains the outcomes of his approach:

---

5   *Las Meninas* in Spanish, generally known in France under the title *Les menines*; Foucault translates it as *Les suivantes*.

Maybe there is in this painting something like a representation of the classical representation and the definition of the space it opens. It effectively undertakes to represent itself in all its elements, with its images, the gazes [*regards*] it offers itself to, the faces it makes visible, the gestures it gives birth to. Yet, in this dispersion it mediates and discloses all together, it necessarily indicates an essential emptiness of all parts: the necessary disappearance of what it is founded upon – the one it resembles and the one in whose eyes it is only resemblance. This subject itself – who is the same – has been elided. Finally freed from the relation it was restrained by, the representation can give itself as pure representation. (Foucault 1966: 31; *).

This analysis was immediately put to the attention of Jacques Lacan who, in 1966, was about to expose his conception of the gaze (*regard*) as 'object A'. In his seminar of 1965-66, following up on his previous year advances on the gaze (*regard*) as *objet petit a*, cause of desire, Lacan insists on the scopic drive (*pulsion scopique*). He brings out its topology, meaning its structure, by making a reference to the projective geometry that, operating in a non Euclidian space, 'allows the reestablishment of the presence of the percipient himself within the field in which he, seemingly unnoticed, is nevertheless perceptible, furthermore adding to the effects of drive [*pulsion*] that manifests itself as exhibition or voyeurism' (Lacan 2005: 326; *).

It is hence in the line of his own research in which Lacan evokes the analysis Foucault undertook about *Les Menines* by Velasquez. As a counterpoint to Foucault's text, Lacan would consecrate three sessions of his seminar to the analysis of the same painting. Psychoanalysis operates a different scene – the analysis of unconscious processes of, for instance, repression – to constitute the perceiving subject (Lacan, 2005). Lacan (28) explains that 'the geometric dimension allows us to see how the subject we are interested in is taken, manoeuvred, and captivated in the visual field'. As a psychoanalyst, Jacques Lacan investigated the original phantasm guiding the hand of the painter, namely the scopic function that Michel Foucault pointed out in his interpretation of the painting. The interpretation of Velasquez's painting by Lacan leads to a quite different interpretation than the one by Foucault. Lacan explains that 'the function of the painter is completely different from the organisation of the representation' (Lacan 2005: 269). In his seminar of 25 May 1966, to make clear what the painter wants to tell us, Lacan returns once again to the analysis of the Velasquez' painting in order. He summarizes his ideas in one sentence: 'You do not see me from where I look at you' (Lacan 2005: 329). This interpretation lends itself to quite different and surely more incisive consequences than those where Foucault's analysis ends.

For those of us working in the specific field of the sociology of international tourism, which has pushed us to interrogate the notion of the tourist gaze, it seems essential to ask the question of the scopic structure that the given-to-see (*le donné-à-voir*) represents. It orients the libido of the subject – and thus to go beyond the denunciation of the representation of the world signified in the given-to-see. The

Figure 14.1   *Peul people, voDaaBé-Bororos, East of Zinder, Niger* by
                  Raymond Depardon

Photo: © Raymond Depardon/Magnum Photos (with the permission of Raymond
Depardon)

analysis of this structure leads us to an interrogation of the relationship between
the *percipens* (the seeing subject) and the *percipi* (what is seen); an interrogation
of the motion of desire that pushes to enjoy (*jouir*) not to enjoy by looking at what
is given to see; in the same way an interrogation of what within the given-to-see
is dissimulated and yet simulated to take hold of the gaze and to mislead it at the
same time.

Analysing a Photograph by Raymond Depardon

The conference in Sheffield from which my present line of questioning had
emerged, had utilized a photograph by Raymond Depardon, *Peul people*, taken
in the East of Zinder, Niger, to communicate the themes discussed during the
event. To conclude, I wish to examine this photograph as in a remarkable way it
condenses the different strings of our foregoing reflections on the gaze. Who is in
the photo? Who is it addressed to? Where has it been taken? Who has taken it?

   As an image, this photograph relates to the imaginary. Its iconic content relates
to the representation of things functioning as a system of signs; signs that according
to Pierce's definition are at the place of another thing for someone (Pierce 1978).

This system of signs is akin of the symbolic order; the real is inscribed with the intentionality of the seeing/seen subject. The real, symbolic and imaginary are articulated within one formulation. We can see a scene of the photographer's life inside an image world. This image world is not the one of the ambient reality. We do not know where exactly this photograph has been taken. All we can tell about the image is that it has been taken in the Sahara (?) desert, amongst 'desert people' figured by their turbaned faces, skilfully disposed in the frame of the window of an immobilized car. There is no decor at the horizon, the background is white. Thus, this image world has its own space and time; the time of the photograph, the time of the taking of the photograph. The image is suspended to the camera; the objective that transforms the instant into a lasting duration beyond any fugitive impression. When I read this photograph as a representative content, when I confer to it any socio-cultural property, I escape from this frame. This photo is not the re-presentation of a landscape; it is essentially a *presentation*. Yet this presentation is a photograph. It relates to a reality that is exterior of itself.

Following Eugen Fink, whose analytical approach I have adopted here, 'the perception of the image is a perceptive mode that has its own original style' (Fink 1966: 89-91). The perception of the image is an experimental mode that constitutes in itself the original place of a factual reality; a place whose unreality is a real appearance but whose reality is suspended in the instant of the photographic being shot, settling the subject in a *durée durante*, to use an expression by Paul Claudel (1946). It is essential to the *imaginous* conscience that we do not dispose of a perceptive piece of data (*donné perceptif*) whose support is hidden (Fink 1966). Can we use this photograph as an empirical support for an analysis of the tourist gaze? There is in this photograph a 'plea to see' (*appel au regard*), a push to see, a frozen picture (*arrêt sur écran*). It can be read both as an operative montage and as a meta-thesis of the visibility of the gaze, to use one of the formulations Michel Foucault uses in his text *Les suivantes* (Foucault [1966] 1998). Following this interpretation, the schemes invented by Lacan can help us to pose sensible, even crucial questions.

In observing the photograph its space to a large part is occupied by the faces and intensive gaze of the 'desert men'. Moreover, on the right, in the restrained frame of the car's rear view mirror, we can see a blurred image which reveals itself as the deformed, reflected image of the photographer himself, while taking the photograph, the objective of his camera pointed to the people that look at him. The strangeness implicit to this play of mirrors confers its singularity and value to this photograph by Raymond Depardon. While looking at the photograph, our own gaze is destabilized. The photograph offers us to look at an ambiguous exchange of gazes (*échange de regards*). The image of the photographer in the car's rear view mirror is a point charged with significance. The subject, presumably representing the tourist gaze, is in the tableau. He is part of the image and, yet, he is dislocated in a twofold way within the space of the photo and the reality it represents. This dislocation introduces a dramatization between the gazer and the gazed-upon, the culmination of an encounter that relates directly to Sartre's thesis of the gaze (*regard*) originating in the existence of the 'other'.

The place where the gazes meet in the tableau does not constitute a moment of pure reciprocity between the seeing and the seen subject. The 'desert men' look at a vanishing point at the horizon symbolizing the cell of the camera. The eye of the photographer is frozen, looking at and capturing the gaze of those he looks upon, while photographing them. He is captured in his own capture. He sees himself seeing. But his image in the car's rear view mirror relates to another place not represented in the picture; a place that exists virtually in the visual field of those who are part of the scene thus making this photograph a highly symbolic illustration of what we can call the 'tourist gaze'. 'In the scopic field, everything is articulated around two terms that work in a dual way. On the one hand there is the gaze, that is, the things that look at me while I see them' (Lacan 1973, 100). At the same time, the gaze of the other that I encounter is not at all a looked-at gaze, but an imagined gaze of the 'other'. It is true that this presence introduces a new message. It accentuates the distance between the 'desert men' and the stranger, between those who possess the (photographic) technology and those who are its target. The camera plays the role of the third gaze that is normally outside the picture and yet is represented in the hands of the photographer. This other gaze fixes their relationship, which, without the camera, would not have this appearance of stability. It is here precisely facing this image that shows us the encounter between figures belonging to different worlds, with the accent put on the relatively exotic character of the respective 'other' (of which the character of strangeness is evoked through the technology used by the photographer), that we experience a feeling of agreement, as if what has been offered for us to see was partly masked. All this makes us become conscious about the incompatibility between the image and the language of the visible.

The photographer is *in* the photo. Has he planned to be in the photo? It is possible, considering Raymond Depardon's experience as a photographer. In any case, the result is there, and as the author of this photograph diffuses it, he will attach a certain value to it. Effectively, the presence of the photographer in the photo, captured at the moment of taking it, confers on this photo an enigmatic character. Yet, it is not the image in the first level of the photograph that attracts our gaze and excites our interest, but the magisterial gesture by which the photographer targets his object and at the same moment includes his signature to the picture. The vision gets dissolved in an instant of seeing. The gaze operates in a certain descent. In Lacan's words (1973: 105); 'it is a question of a kind of desire at [*à*] the 'other' leading to the given-to see'.

I recently listened to a French radio interview with Raymond Depardon during which he explained that when he was a child he was a little timid and didn't dare to look at strangers, and then went on to become a photographer. From this point of view, his photo could be considered as a double act; monstration and, at the same time, annulation of this monstration perceived by the subject as a separation between himself and the stranger.

References

Bachelard, G. (1949), *La philosophie du Non. Essai d'une philosophie du Nouvel Esprit scientifique*, PUF, Paris.

Bachelard, G. (1985), *The New Scientific Spirit*, Beacon Press, London.

Baudrillard, J. (1970), *La société de consommation, ses mythes, ses structures*, Le point de la question, SGPP, Paris.

Baudrillard, J. (1972). *Pour une critique de l'économie politique du signe*, Gallimard, Paris.

Baudrillard, J. (1981 [1973]), *For a Critique of the Political Economy of the Sign*, Telos Press, St. Louis.

Berkeley, G. (1732), *An Essay Towards a New Theory of Vision*, 4th edn, <http://psychclassics.yorku.ca/Berkeley/vision.htm> retrieved 8 April 2008.

Boorstin D.J. (1961), *The Image: A Guide to Pseudo-Events in America*, Harper & Row, New York.

Canguilhem G. (1977), *Ideologie et Rationalité dans l'histoire des Sciences de la vie*, Vrin, Paris.

Canguilhem, G. (1991), *The Normal and the Pathological*, transl. by C.R. Fawcett and R.S. Cohen, Zone Books, New York.

Claudel, P. (1946), 'Les psaumes et la photographie', in his *L'oeil écoute*, Gallimard, Paris.

Debord, G. (1983), *The Society of Spectacle*, Black & Red, Detroit, MI.

Deleuze, G. (1986), *Foucault*, Minuit, Paris.

Fink, E. (1966), *De la phénoménologie*, Minuit, Paris.

Foucault, M. (1963), *Naissance de la clinique: Archéologie du regard medical*, PUF, Paris.

Foucault, M. (1971), *L'ordre du discours – Leçon inaugurale au Collège de France*, Flammarion, Paris.

Foucault, M. (1998 [1966]), *Les mots et les choses, Archéologie du Savoir*, Gallimard, Paris.

Foucault, M. (2000), *Naissance de la clinique*, 6ᵉ ed., PUF, Paris.

Foucault, M. (2001), *The Order of Things: Archaeology of the Human Sciences*, Routledge, London.

Foucault, M. (2002), *Archaeology of Knowledge*, Routledge, London.

Foucault, M. (2005), *History of Madness*, transl. by J. Khalfa and J. Murphy, Routledge, London.

Heidegger, M. (1927), *Sein und Zeit*, Niemeyer, Tübingen.

Heidegger, M. (1962), *Time and Being*, Blackwell, Oxford.

Heidegger, M. (1986), *L'Etre et le Temps*, 2ᵉ ed., Gallimard, Paris.

Hollinshead, K. (2000), 'The Tourist Gaze and its Games of Truth: An elaboration of 'the governmentality' of Foucault via Urry', in M. Robinson (ed.) *Motivations, Behaviour and Tourist Types*, Business Education Publishers, Sunderland.

Husserl, E. (1982) *Recherches phénoménologiques pour la constitution*, PUF, Paris.

Husserl, E. (1985 [1970]), 'L'idée de la phénoménologie', in his *Cinq leçons prononcées en 1907 à l'Université de Gottingen en introduction à un cours sur la constitution des choses spatio-temporelles*, PUF, Paris.

Husserl, E. (1999 [1950]) *Cartesian Meditations: An Introduction to Phenomenology*, transl. by D. Cairns. Kluver, Dordrecht and London.

Lacan, J. (1966), 'Le stade du miroir comme formateur de la fonction du Je telle qu'elle nous est révélée dans l'expérience psychanalytique', in his *Ecrits*, Seuil, Paris, pp. 93-100.

Lacan, J. (1973), *Le Séminaire 1964/65: les quatre concepts fondamentaux de la psychanalyse*, Seuil, Paris.

Lacan, J. (1982), *Ecrits: a Selection*, W.W. Norton and Co. Inc, New York.

Lacan, J. (1998), *The Four Fundamental Concepts of Psycho-Analysis*, Vintage, London.

Lacan, J. (2005), *L'objet de la psychanalyse, séminaire 1965/66*, édition ALI, Hors commerce.

Lanfant, M.-F. (1992), 'Le tourisme international entre tradition et modernité, Pourquoi ce thème?', *Actes du colloque*, Université de Nice/URESTI-CNRS, Paris, pp. 1-25.

Lanfant, M.-F., Allcock, J.B. and Bruner E.M. (eds) (1995), *Tourisme International Identity and Change*, Sage Publications, London.

MacCannell, D. (1976), *The Tourist, A New Theory of the Leisure Class*, Schocken Books, New York.

Merleau-Ponty, M. (1964), *Le visible et l'invisible*, ed. by C. Lefort, Gallimard, Paris.

Merleau-Ponty, M, (1968), *Résumés de cours: Collège de France 1952-1960*, Gallimard, Paris.

Merleau-Ponty, M. (1945), *La phénoménologie de la perception*, Gallimard, Paris.

Merleau-Ponty, M. (1960), *Signes*, Gallimard, Paris.

Merleau-Ponty, M. (1962), *Phenomenology of Perception*, Kegan Paul, London.

Merleau-Ponty, M. (1964), *Signs*, Northwestern University Press.

Merleau-Ponty, M. (1969), *The Visible and the Invisible*, Northwestern University Press, Evanston.

Pierce, C.S. (1978), *Écrits: 1903-1910*, transl. and ed. by G. Deledalle, Seuil, Paris.

Robinson, M. and Picard, D. (2003), *Tourism and Photography: Still Visions, Changing Lives*, Conference Proceedings, CTCC, Leeds Metropolitan University, Leeds.

Sartre, J.-P. (1943), *L'être et le néant*, Gallimard, Paris.

Sartre, J.-P. ([1957] 2007), *Being and Nothingness: An Essay on Phenomenological Ontology*, Routledge, London.

Urry, J. (1990), *The Tourist Gaze*, Sage, London.

Urry, J. (2002), *The Tourist Gaze*, 2nd edn, Sage, London.

Index